ANALYSIS AND SOLUTIONS FOR SWITCHING NOISE COUPLING IN MIXED-SIGNAL ICs

ANALYSIS AND SOLUTIONS FOR SWITCHING NOISE COUPLING IN MIXED-SIGNAL ICs

by

Xavier Aragonès
Universitat Politècnica de Catalunya (UPC)

José Luis González
Universitat Politècnica de Catalunya (UPC)

and

Antonio Rubio
Universitat Politècnica de Catalunya (UPC)

Springer-Science+Business Media, B.V.

A C.I.P. Catalogue record for this book is available from the Library of Congress.

ISBN 978-1-4419-5085-7 ISBN 978-1-4757-3013-5 (eBook)
DOI 10.1007/978-1-4757-3013-5

Printed on acid-free paper

Contents

1 Introduction

Today's VLSI electronic technology allows complex mixed-mode circuits to be integrated in a single silicon die, with high speed digital circuits fitted alongside high performance analog sections. Noise immunity has been one of the classical drawbacks of analog circuits as opposed to digital ones. In the last few years, the resolution demanded of such circuits has increased because of the need to process very low level signals and through the increasing quality demanded in audio or video communications. Analog circuit designers traditionally had to deal with noise sources related with devices, mainly thermal, shot and flicker (1/f) noise [1]. The trends in microelectronics are nevertheless heading towards integrating full systems in a single chip, including analog and digital parts. This fact makes device noise no longer the main source of noise affecting analog circuitry, this being instead the disturbances that digital circuitry induces over the analog part. The problem is made worse because of the increasing density of integration and the growing speed of digital circuits, which make the noise produced larger and larger.

Coupling problems have become a limiting factor in the performance of many advanced circuits. This has brought about a need for research in the field, in order to find the way all these effects are produced, and what can be done to avoid them. This book focuses on the analysis of coupling mechanisms, specifically noise coupled through the silicon substrate shared by analog and digital circuitry in a mixed-signal IC due to dV/dt *as well as on the parasitic coupling through packaging due to* dI/dt. *This introductory chapter contains a review of the state of the art in microelectronics, to look at its evolution and thus understand the factors which lead to the aforementioned situation*

1.1 Trends in microelectronic technology

It is widely known that one of the aspects characterizing microelectronics is its constant trend towards continuous miniaturization. The reduction in size of the elements forming an IC is known as *scaling down*, and the classic trend has been to halve the device size every two generations of technology –six years–, i.e. the number of devices scaled by a factor of four [2]. The dimensions of the elements in a circuit are usually related to a reference size (feature size) characteristic of each technology. Figure 1.1 shows the recent evolution of this feature size –and consequently the evolution in the sizes of all the elements in the circuit–, revealing the miniaturization trend stated above.

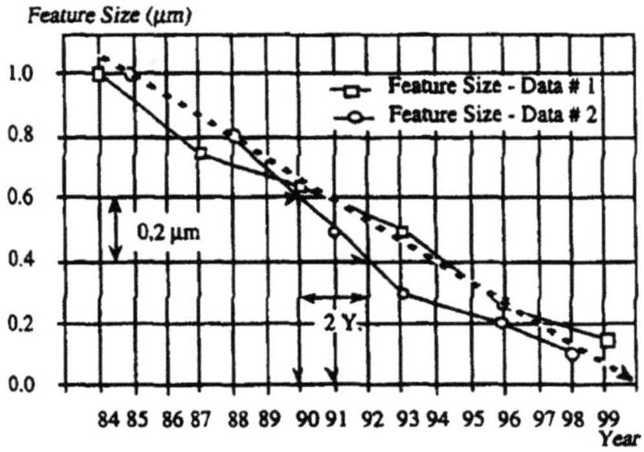

Figure 1.1

Device feature size evolution over the last few years [3].

If electric fields are to be maintained in a device, not only physical dimension, but also the supply voltage must be scaled, while impurity concentration must increase by the inverse of the scaling factor. This scaling scenario reducing both geometry and voltages by a factor α makes power dissipation drop by a factor α^2, integration density also increase by a factor α^2, and speed multiply by a factor α [4]. Although all these consequences are positive and produce an improvement in circuit performance, they also imply more noise problems. A coupling between two elements in a circuit will be produced through some parasitic element: a resistance, a capacitance, or an inductance. Greater integration density implies smaller distances between circuit elements. If coupling is produced through a parasitic resistance, less distance implies lower attenuation of the disturbance. If the parasitic element is a capacitance, less distance means higher capacitance, and more coupling. On the other hand, coupling through capacitive or inductive elements is proportional to the rate of voltage or current variation, respectively. More speed thus means more coupling through parasitic capacitances or inductances. In short, the trend towards miniaturization leads to an increase in couplings between the elements of an IC.

In practice, supply voltage has not been scaled the same way as geometry has, due to the advantage of having one standard supply level, and also to the leakage current increase produced when scaling down transistor threshold voltage. Non-scaled voltages have produced nothing but an even larger increase in speed. Nevertheless, reliability issues have finally compelled voltage-scaling [2]. On the other hand, scaling down both interconnect width and thickness lead current density to increase by a factor α^2, while the delay due to interconnects remains constant. If interconnect width is not scaled, integration density is lost and delays increase (for a given length). Keeping the thickness of metal layers constant or even increasing it has been proposed as an alternative, but implies more capacitance between neighboring interconnects, and thus more coupling problems. Scaling down the elements in an IC has not led to chip size reduction, but market laws have favored a rush for better performance and consequently higher circuit complexity. In 1975 the well known Moore's Law was stated [5], according to which integration density doubles every 18 months. This law has been accomplished since then [6]. Figure 1.2 shows the evolution of the number of transistors per chip over the last few years, for microprocessors and memories, and how this evolution has led to larger die sizes.

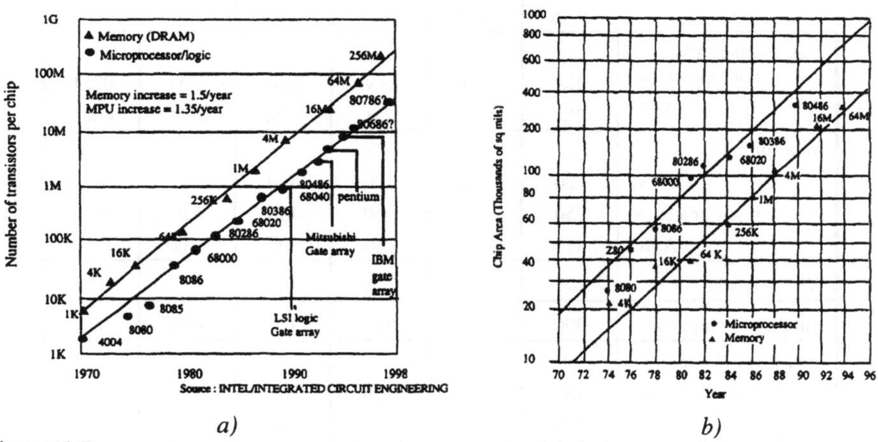

Figure 1.2

Evolution of the complexity (a) and die size (b) of ICs [3].

In fact, increasing the complexity of circuits not only leads to larger chips, but also to an increase in the number of inputs and outputs –package pins–, number of power supply terminals, number of metal layers for internal wiring, the length of the interconnects, and package dimensions. Table 1.I shows forecasts for the characteristics of ICs in the near future, and it can be foreseen that in a few years' time integrated circuits will be expected to have seven wiring layers, pin count will be in the thousands and speeds will soar over GHz frequencies.

All these trends also have their effect on couplings. The increasing number of wiring levels clearly eases interaction between different signals. The larger number of inputs and outputs is also a problem, given the high driving capacity of the corresponding pads, and the possibility to have many of those pads switching simultaneously. Similarly, increasing the complexity of sequential CMOS circuits means power consumption being concentrated at determined points of time. Apart from the reliability and power dissipation problems this may cause, large current peaks mean large amounts of noise due to the parasitic inductive elements.

Table 1.I.

Forecasts for microelectronics characteristics according to the Semiconductor Industry Association (SIA) [7].

	1995	1998	2001	2004	2007	2010
Lithography (μm)	0.35	0.25	0.18	0.12	0.1	0.07
Transistors/cm^2:						
microprocessor	4M	7M	13M	25M	50M	90M
ASIC	2M	4M	7M	12M	25M	40M
Area (mm^2):						
DRAM	190	280	420	640	960	1400
microprocessor	250	300	360	430	520	620
ASIC	450	660	750	900	1100	1400
Wiring levels	4-5	5	5-6	6	6-7	7-8
Number of chip I/Os	900	1350	2000	2600	3600	4800
Frequency (MHz):						
chip-to-board	150	200	250	300	375	475
on-chip	300	450	600	800	1000	1100

Circuit packaging is also an interesting issue. The higher number of pins leads to larger package dimensions, and this always involves an increase in parasitic pin inductances. Package inductances are responsible for an important source of noise, as will be seen below, and minimizing them would be desirable. On the other hand, the classic bonding approach is not feasible when there is a very large pin count. Power supply and I/O pads are conventionally aligned at the chip periphery, and they are wire-bonded to the package leadframe. If the number of terminals is high enough, this leads to layouts in which the area is determined by the rectangle formed by the pads – pad limited designs–, much larger than the area of the internal circuitry –the core–. New types of packages are therefore needed, in which bonding is performed directly on the chip surface, avoiding the need to place the pads on the edges –for example, flip-chip technologies [8]–. Studies will be needed to find out the possibility of interference between package leads and chip wiring, given their proximity.

Apart from the increasing complexity of ICs, there is a parallel trend (more difficult to quantify) towards integrating different technologies to develop high performance circuits. The first step in this direction was made by BiMOS technologies, appearing in the mid 80's, in which both MOS and bipolar transistors were integrated in the same die, by adding several steps to a typical MOS process. On the other hand, the higher density and flexibility of integration has led to increasingly complex circuits, up to the present possibility of having full systems in a single die. Particularly in telecommunications circuits, this means integrating analog and digital circuitry in the same chip, resulting in what is known as mixed-signal circuits. This has given rise to the appearance of serious electric coupling problems, because of the noisy nature of digital circuits –fast voltage and current transients– and the low noise tolerance of analog circuits. The increasing demand for resolution has merely turned insulation between analog and digital parts into a major problem, so much so that many applications could not reach the desired low noise levels, thus limiting their performance.

In order to minimize coupling and packaging problems, alternative technologies like multichip modules (MCM) have been put forward. To put it succinctly, these technologies consist of bonding different ICs or lumped devices to a substrate making connection tasks. This substrate is typically formed by a number of interconnect layers with dielectrics in between, while ICs are not packaged and directly bonded to the substrate. This allows reducing the package pin count, especially connected with the circuit complexity, and on the other hand allows having different technologies like MOS, bipolar or GaAs in the same package. Multichip modules enable splitting in different die circuits with coupling problems, while maintaining the whole set inside the same package. Nevertheless, this MCM technology has a significantly higher cost than a monolithic IC, and is thus not yet an attractive alternative to mixed signal circuits.

Apart from the possibility of integrating both MOS and bipolar technologies, and analog and digital sections, the race does not stop there, and many modern processes allow high voltage devices –typically between 30 and 50 V– to be implemented together with conventional circuitry. These smart power technologies no doubt take the appearance of disturbances, though high power and high resolution are not usually demanded at the same time.

Once the IC is manufactured, it must be packaged. We have already discussed the bonding of input/output signals and power supplies, but this wire-bonding does not provide a strong enough join between chip and package. The chip is usually stuck with epoxy to a metal backplane in the package. This epoxy may be designed to provide high thermal conductivity, or high electrical conductivity. A metal layer may be deposited at the die backside, thus creating a conductive backplane short-circuiting the silicon bulk and the package metal plane. This backplane may then be used to bias the substrate wire-bonding the package metal plane to the ground pins.

1.2 Types of parasitic coupling in mixed-signal circuits

The former section showed how the trend towards greater complexity, increasing performance and technology scaling makes couplings more important. In applications where disturbances can limit circuit performance, they should be taken into account in all design stages. The main mechanisms of electric coupling in ICs are briefly presented below. When the parasitic coupling is capacitive the noise is caused by *dV/dt* of the signals, and this is caused by *dI/dt* when the coupling is inductive.

1.2.1. Noise coupled between interconnects (dV/dt)

Capacitive coupling between interconnects, traditionally simply known as crosstalk, has been studied since the 80's [9], [10]. The effect is due to the parasitic capacitance between any two parallel interconnects. A voltage transient in one of the interconnects produces a current flow through the capacitance, and a noise peak at the neighbor line –positive peak if a rising transient, negative peak for a falling transient–. The larger the capacitance the larger the coupling, which means it increases for closer and longer interconnects. Another effect capacitive coupling may have is a variation –upward or downward– of the delay of a signal at the affected line, which may produce set-up or hold time violations in synchronous memory elements [11].

Crosstalk phenomena were already a problem in printed circuit boards, and the arrival of high-speed LSI has made this a center of attention in digital circuits because of the high noise levèls produced –around some volts–. Spurious signals may propagate indefinitely through logic gates, depending also on the pulse width, and produce permanent errors if they affect memory elements [12]. In mixed-signal circuits crosstalk is not usually a limiting factor, since care is generally taken to split digital and analog domains when floorplanning the IC, avoiding proximity between digital and analog interconnects. Nevertheless, some blocks need both kinds of signals –a sample-and-hold, for instance–,demanding a very careful signal distribution, moving digital interconnects away from sensitive nodes. If necessary, coupled lines may be shielded through the insertion of a third grounded line in between, so noise is reduced.

In the last few years we have seen the emergence of CAD tools which take into account capacitive coupling in the routing generation, some of these having the ability to perform modifications –increase distance, add shielding– in order to eliminate crosstalk in ICs or in multichip modules [13].

1.2.2. Noise coupled through the substrate (dV/dt)

It has traditionally been considered that a silicon substrate provides satisfactory insulation between devices, because of its high resistivity and the fact that any junction to the substrate is reverse-biased. Nevertheless, there are some mechanisms through which switching devices can introduce disturbances to the substrate that will propagate to other devices due to its finite resistivity. Figure 1.3 shows the common technological characteristics of two typical types of substrates. Wafers with a uniform doping from surface to bottom are called *low conductive*, *lightly doped*, or simply *P-*. To minimize latch-up problems, a very low resistive wafer is used, in which a few micron thick layer with the usual conductivity is grown by means of epitaxy. This kind of wafer is known as *high conductive*, *heavily doped*, or simply *P+* (because of their doping level). The high resistivity fosters noise attenuation, so the amount of noise that reaches the sensitive circuits is not very large, but anyway the increasing demand for resolution in analog applications has made this coupling mechanism a limiting factor in the performance of many mixed-signal circuits [14]. Furthermore, the usage of P+ wafers favors noise conduction towards remote parts of the IC. Lately it has also been reported that substrate coupling demands an increase in the refreshment frequency in dynamic RAM memories [15]. Going by intuition and experience, some techniques for minimizing substrate coupling are being implemented, such as increasing distance between coupled circuitry, using different types of guard rings, wells, buried layers, increasing the number of substrate contacts, etc. [16]-[17]. There are no clear criteria about the effectiveness of these techniques, because the first analyses on the subject were not published until 1992 [18].

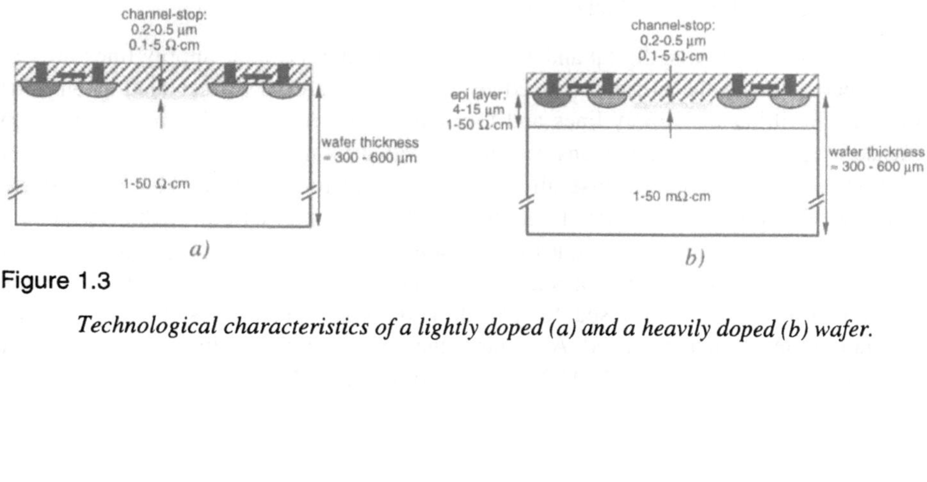

Figure 1.3

Technological characteristics of a lightly doped (a) and a heavily doped (b) wafer.

A different kind of wafer that provides better isolation is that of silicon-on-insulator (SOI) technologies. Either by implant and oxidation, either by deposition, a thick insulator layer –typically silicon oxide– is created some microns under the wafer surface. The different parts of the circuit –even single devices– may be isolated by oxide trenches, thus eliminating any conductive path between them.

1.2.3. Noise coupled through power-supply lines (dI/dt)

The set of wiring connecting a chip pad to the board can be modeled as an inductance in parallel with a capacitance [19]. In power-supply pins, where voltages are approximately constant, the inductance is the dominant element. In CMOS circuits, where consumption is not steady, current peaks produce voltage drops through the inductance, which manifest themselves as fluctuations in the internal supply voltages. This kind of noise is known as delta-I noise, simultaneous switching noise (SSN) or ground bounce. The most important sources of SSN in digital circuits are output pads, because of their driving capabilities. An usual expression to evaluate the amount of noise in a supply line produced by n simultaneously switching output nodes is

$$V_n = nL_{ef}\frac{dI}{dt} \qquad \text{Eq. (1.1)}$$

where L_{ef} is the effective pin inductance of the corresponding line –see Figure 1.4–. This expression overestimates noise [20], but it is useful to know that noise increases with the number of gates demanding current at a given time, and with the effective inductance of the power-supply pins. This last dependency indicates the importance of the characteristics of the package and of pin assignment, because those closer to the package corners are the most inductive ones [21]. Multiple pin assignment for the power supplies, and especially thick connections –lug pins– are recommendations made to reduce effective inductance and thus noise at the supply lines.

In mixed-signal circuits, digital and analog parts sharing the same supply lines provide a perfect path for the noise produced by switching gates to affect analog circuitry directly. Splitting the supply lines and terminals of each part is the most elementary measure taken to obtain better insulation. Anyway a total independence between both sections is not possible because all grounds are usually connected to the substrate, when not the foundry itself, which requires short-circuiting them through a scribe lane at the periphery. Several techniques for reducing SSN have been proposed in the last years, with different degrees of success. Concerning the noise caused by the pads, techniques have been proposed for avoiding their simultaneous switching, and for regulating their switching speed. As regards the noise originating in the internal digital circuitry, certain current-mode MOS logic families have been proposed –ESCL [22],

FSCL [23]– which get almost constant power consumption using structures analog to ECL gates in the bipolar world. Their main drawback is the large quiescent consumption, which is excessive compared to conventional CMOS logic. Last, noise reduction can be achieved with on-chip decoupling capacitors, or off-chip devices but inside the same package, or by other more imaginative proposals like current feeding by internal tank capacitors.

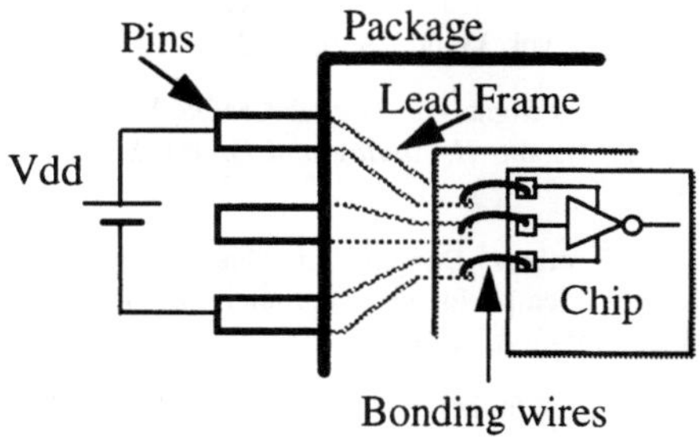

Figure 1.4

Schematic illustration of the package structure of an integrated circuit showing the main parts that cause packaging inductance.

References

[1] *Low-Noise Wide-Band Amplifiers in Bipolar and CMOS Technologies*, Z.Y. Chang, W. Sansen, Chapt. 2, Kluwer Academic Publishers, 1991.

[2] Chenming Hu, "Future CMOS Scaling and Reliability", *Proceedings of the IEEE*, vol. 81, no. 5, pp. 682-689, May 1993.

[3] *CAD and Testing of ICs and Systems. Where are we going?*, B. Courtois, TIMA-CMP, France 1996.

[4] R.H. Dennard, F.H. Gaensslen, H.N. Yu, V.L. Rideout, E. Bassous, A.R. LeBlanc, "Design of Ion-Implanted MOSFET's with Very Small Physical Dimensions", *IEEE J. Solid-State Circuits*, vol. 9, no. 5, pp. 256-267, October 1974.

[5] G.E. Moore, "Progress in Digital Integrated Electronics", *Proc. IEEE Int. Electron Devices Meeting*, pp. 11-13, 1975.

[6] G.E. Moore, "Lithography and the Future of Moore's Law", *Optical/Laser Microlithography VIII: Proceedings of the SPIE*, 2440, pp. 2-17, February 1995.

[7] *The National Technology Roadmap for Semiconductors*, Semiconductor Industry Association, 1994, document available at: http://www.sematech.org/public/roadmap/index.htm.

[8] *Physical Architecture of VLSI Systems*, R.J. Hannemann, A.D. Kraus, M. Precht, John Wiley & Sons, 1994.

[9] E.T. Lewis, "An Analysis of Interconnect Line Capacitance and Coupling for VLSI Circuits", *Solid-State Electronics*, vol. 27, nos. 8/9, pp. 741-749, 1984.

[10] S. Seki, H. Hasegawa, "Analysis of Crosstalk in Very High-Speed LSI/VLSI's Using a Coupled Multiconductor MIS Microstrip Line Model", *IEEE Trans. Microwave Theory and Techniques*, vol. 32, no. 12, pp. 1715-1720, December 1984.

[11] F. Moll, M. Roca, A. Rubio, "Measurement of Crosstalk-Induced Delay Errors in Integrated Circuits", *Electronics Letters*, vol. 33, no. 19, pp. 1623-1624, September 1997.

[12] F. Moll, A. Rubio, "Spurious Signals in Digital CMOS VLSI Circuits, a Propagation Analysis", *IEEE Trans. Circuits and Systems II: Analog and Digital Signal Processing*, vol. 39, no. 10, pp. 749-752, October 1992.

[13] T. Hameenanttila, J.D. Carothers, Donghui Li, "Fast Coupled Noise Estimation for Crosstalk Avoidance in the MCG Multichip Module Autorouter", *IEEE Trans. Very Large Scale Integration Systems*, vol. 4, no. 3, pp. 356-368, September 1996.

[14] B.P. Brandt, B.A. Wooley, "A 50-MHz Multibit Sigma-Delta Modulator for 12-b 2-MHz A/D Conversion", *IEEE J. Solid-State Circuits*, vol. 26, no. 12, pp. 1746-1756, December 1991.

[15] T. Tsuruda, M. Kobayashi, M. Tsukude, T. Yamagata, K. Arimoto, M. Yamada, "High-Speed/High Bandwidth Design Methodologies for On-Chip DRAM Core Multimedia System LSI's", *IEEE J. Solid-State Circuits*, vol. 32, no. 3, pp. 477-482, March 1997.

[16] E.A. Vittoz, "The Design of High-Performance Analog Circuits on Digital CMOS Chips", *IEEE J. Solid-State Circuits*, vol. 20, no. 3, pp. 655-665, June 1985.

[17] D.K. Su, B.A. Wooley, "A CMOS Oversampling D/A Converter with a Current-Mode Semidigital Reconstruction Filter", *IEEE J. Solid-State Circuits*, vol. 28, no. 12, pp. 1224-1233, December 1993.

[18] D.K. Su, M.J. Loinaz, B.A. Wooley, "Experimental Results and Modeling Techniques for Switching Noise in Mixed-Signal Integrated Circuits", *Proc. Symposium on VLSI Circuits*, pp. 42-43, 1992.

[19] R. Senthinathan, J.L. Prince, "Simultaneous Switching Ground Noise Calculation for Packaged CMOS Devices", *IEEE J. Solid-State Circuits*, vol. 26, no. 11, pp. 1724-1728, November 1991.

[20] *Simultaneous Switching Noise of CMOS Devices and Systems*, R. Senthinathan, J.L. Prince, Kluwer Academic Publishers, 1994.

[21] J.M. Williamson, M.S. Nakhla, Q. Zhang, P.D. van der Puije, "Ground Noise Minimization in Integrated Circuit Packages through Pin Assignment Optimization", *IEEE Trans. Components, Packaging and Manufacturing - Part B*, vol. 19, no. 2, pp. 361-367, May 1996.

[22] M. Maleki, S. Kiaei, "Enhancement Source-Coupled Logic for Mixed-Signal VLSI Circuits", *IEEE Trans. Circuits and Systems II: Analog and Digital Signal Processing*, vol. 39, no. 6, pp. 399-402, June 1992.

[23] D.J. Allstot, S. Chee, S. Kiaei, M. Shrivastawa, "Folded Source-Coupled Logic vs. CMOS Static Logic for Low-Noise Mixed-Signal ICs", *IEEE Trans. Circuits and Systems I: Fundamental Theory and Applications*, vol. 40, no. 9, pp. 553-563, September 1993.

2 Substrate noise characteristics and propagation

This chapter contains an exhaustive attempt to reveal the elementary characteristics of substrate coupling. The aim of this part is to determine how coupling is affected by factors such as the type of wafer, the type of coupled devices, the biasing of these devices, and layout and technology parameters.

The chapter starts with a description of the mechanisms by which circuit nodes interact with the substrate. Next, the way device characteristics –layout and technology parameters, device type and biasing– affect coupling is studied with device simulations and experimental measurements. Lastly, we analyze the way noise propagation is affected by the substrate type or particular biasing, assuming ideal grounding conditions. In following chapters the consequences of non-ideal supplies on the results obtained will be analyzed.

2.1 Noise injection and reception mechanisms

Any switching node can couple noise to the substrate, whose finite resistivity means that such noise may be transmitted to neighboring devices. There are several mechanisms by which noise is injected and received, and these should be identified in order to improve our understanding of how effective steps that may be taken to eliminate the noise trouble are, and their importance in every kind of device.

Figure 2.1 shows an illustration of the main noise injection and reception mechanisms, which are now looked at in detail.

2.1.1. Substrate noise injection mechanisms

2.1.1.a. Capacitive injection through reverse-biased junctions.
The most evident mechanism of substrate noise injection is coupling across the PN junctions formed between devices and the substrate or well. If devices are operated

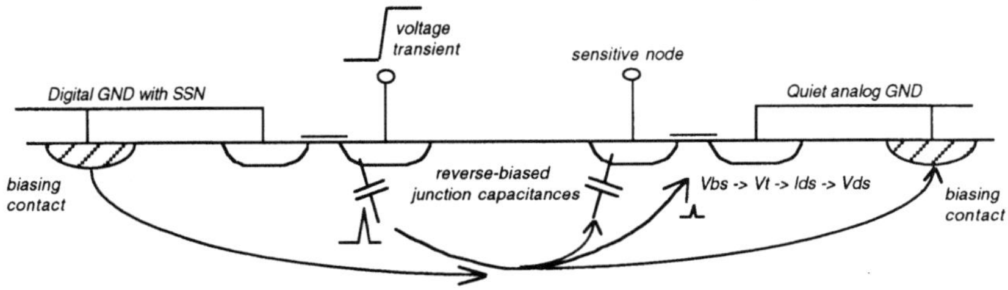

Figure 2.1

Noise is mainly injected through switching device junctions, and substrate contacts connected to noisy lines –left–, and received by sensitive devices through the junctions, through body effect, and through the power-supply connected to the substrate –right–.

normally, junctions are reverse-biased and their AC behavior is like that of a capacitance, valued as

$$C_j = \frac{A\varepsilon_{Si}}{\left[\frac{2\varepsilon_{Si}}{q} V_{bi}\left(\frac{1}{N_A} + \frac{1}{N_D}\right)\right]^{1/2} \cdot \left(1 - \frac{V_A}{V_{bi}}\right)^m} \qquad Eq.\ (2.1)$$

where A is the junction area, V_{bi} is the built-in junction voltage, V_A the voltage difference applied, N_A and N_D the dopings at regions P and N, respectively, and m is a fitting factor, which for a steep junction takes value $1/2$ [1]. Given that one of the sides of the junction –substrate or well– is at a constant voltage, current flows through the capacitance when a transient voltage is produced at the device side. This current is proportional to the transition speed and capacitance area, so noise injection will be important in high-speed digital devices, and in large drivers.

In MOS devices, noisy nodes are drains in digital gates, especially those of output buffers. Noise injection by devices inside a well (PMOS devices in N-well technologies) is less important if the sensitive device is in a different well. In bipolar vertical transistors, the node with the highest noise injection capability is the collector, as will be seen in section 2.3, because it is the node forming a PN junction with the substrate. The emitter and the base are more insulated from the substrate, and are therefore less troublesome. In lateral transistors, the most noisy node is the base. Other devices that may form junctions with the substrate are diffused resistances. Nevertheless, these elements are mainly used in analog circuits, thus being noise sensitive elements rather than noise generators.

2.1.1.b. Noise injection through contacts.

Power-supply lines in digital circuits are usually strongly contaminated with noise, given the package impedance to the voltage sources[1]. This noise, known as simultaneous switching noise, may easily reach hundreds of millivolts [2]. If these noisy lines are used to bias the substrate, the ohmic junction of the contact allows noise injection. Moreover, as contacts are placed all along the chip surface, noise levels will be approximately uniform in the substrate, unlike the situation in which noise is injected at a point. As will be seen in chapter 4, this noise source can easily become more important than noise injection through device junctions. The biasing strategies allowing this mechanism to be minimized will be analyzed in detail there, together with the package influence.

2.1.1.c. Other mechanisms.

The parasitic capacitance between an interconnect and the substrate, mainly if the former is on the lowest metal layer, may ease coupling when the interconnect drives a digital signal. This noise is in fact seen as a superposition to the noise capacitively injected through the junctions in the interconnect driver. If the interconnect length is a few tens of microns, its substrate capacitance is already comparable to the drain junction capacitance of a minimum-size transistor, and the noise injected is comparable. A particular case of coupling between metal layers and substrate is the sort produced by integrated inductors in RF circuits [3], [4].

A source of coupling mentioned in [5] is the forward biasing of device junctions. The aforementioned work analyzes a mixed-signal circuit presenting noise problems, finding that the noise present at the digital ground line forward biases the protection diodes in output pads, which allows injection of up to 200 mV of substrate noise.

Lastly, noise injection due to hot carriers deserves more detailed analysis. It is known that the high electric field between drain and source in submicron transistors, makes electrons –in NMOS devices– acquire enough energy to become what is called *hot*. These hot electrons impact the drain, producing ionization and dislodging holes that are displaced to the substrate, appearing as a substrate current [6]. This impact ionization current dominates junction leakage, reaching typical levels of microamps per micron [7], [8]. Hot carrier effects have been observed even in low-voltage devices [9]. In PMOS transistors, hot carrier substrate currents are smaller, due to holes lower mobility.

Among the multiple effects that impact ionization currents may produce, appears the disturbance of other circuits in the same IC. In particular, in [10] it is stated that, for low frequencies (< 100 MHz) impact ionization current is the most important of substrate coupling sources, followed by the noise introduced by the contacts, and

[1] Throughout this work we will assume chip terminals directly connected to voltage sources, i.e. we will disregard noise at board level.

capacitive coupling through the junctions. The authors set up a test circuit with a noisy section and an A/D converter acting as a sensor, both sections being completely independent except for the common substrate. The circuit allows the substrate contacts to be connected to the noisy MOS sources. The measurement results do not display any differences between both biasing options, and the authors conclude that the main noise source is not substrate contacts, but devices, and they assume that the main coupling mechanism is impact ionization. Nevertheless, they ignore capacitive coupling through junctions, which might well be the main coupling mechanism in this case.

In order to determine the relative importance of impact ionization as compared with purely capacitive coupling, the device simulator MEDICI [11] is used to evaluate the amount of current injected in the substrate due to both mechanisms. A two-dimensional simulation of two close NMOS transistors is performed, with one of the transistors switching. Devices are defined using 0.7 μm technology data, and two simulations are made, one in which MEDICI considers impact ionization, and the other one ignoring it. Figure 2.2 shows the current density in the substrate in a cross-section 1 μm deep, under the transistors. The drain of the transistor –switching from 5 to 0 volts– is placed between microns 6 and 8 in the horizontal axis, while the drain of the sensitive device is placed between microns 24 and 26. It can be observed that the contribution of substrate current due to hot carriers is similar to what comes from junction capacitances –a few microamps per micron. Simulations with data taken from a 1.5 μm technology have been also performed –doubling channel length–, and the contribution of impact ionization is then much lower, roughly 20% of the total substrate current. It is therefore concluded that it is not true that the current produced by hot carriers is dominant, but its importance increases with technology scaling down, and it is not until present technologies that it becomes comparable with the current injected capacitively.

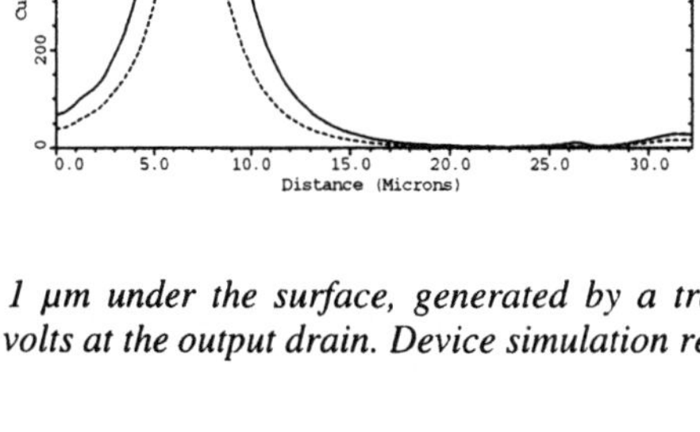

Figure 2.2

Current density 1 μm under the surface, generated by a transistor with L=0.7 μm, transition 5 to 0 volts at the output drain. Device simulation results.

2.1.2. Noise reception mechanisms

2.1.2.a. Capacitive reception through reverse-biased junctions.
Just as a junction capacitance allows disturbance injection when drain voltage switches, this capacitance also allows noise reception when there is a voltage variation between the substrate and the sensitive node, typically the drain. Noise can also be coupled to the channel through depletion regions, although this can be seen as an extra contribution to the noise injected to the source or drain.

The capacitive mechanism is also especially important in passive devices implemented on the substrate (resistances, capacitances), and in lateral bipolar transistors, because then noise couples directly to the base, and is amplified.

2.1.2.b. Body effect.
It is well known that threshold voltage depends on bulk-source voltage, V_{bs}. For a NMOS transistor, this dependence is given by

$$V_T = V_{TO} + \gamma\left(\sqrt{2\phi_b - V_{BS}} - \sqrt{2\phi_b}\right) \qquad \textit{Eq. (2.2)}$$

where γ is a substrate threshold parameter, and ϕ_b is the surface potential in strong inversion. These two parameters can be expressed as

$$\gamma = \frac{\sqrt{2q\varepsilon_{Si} N_A}}{C_{ox}} \qquad \textit{Eq. (2.3)}$$

$$\phi_b = \frac{KT}{q}\ln\left(\frac{N_A}{n_i}\right) \qquad \textit{Eq. (2.4)}$$

When a disturbance appears at the substrate, voltage V_{bs} varies, and so does the threshold voltage V_T, the drain current I_d, and the drain voltage V_{ds}. This phenomenon is called the body effect. A transient substrate voltage increase is displayed as a negative drain noise peak, and vice-versa. The magnitude of the noise acquired through the body effect is similar to the noise acquired capacitively. Figure 2.3 represents the dependence of the noise acquired through this mechanism on the substrate doping, for a saturated transistor with V_{ds}=2.5 V, substrate noise level V_{bs}=10 mV, and the remaining parameters taken from a typical 1.5 μm technology. It is verified that as substrate doping increases, the disturbance at the sensitive device also increases. One should remember that the scaling laws lead to a continuous increase in doping levels. Lastly observe that this mechanism is not dependent on the speed, unlike coupling through the device junctions.

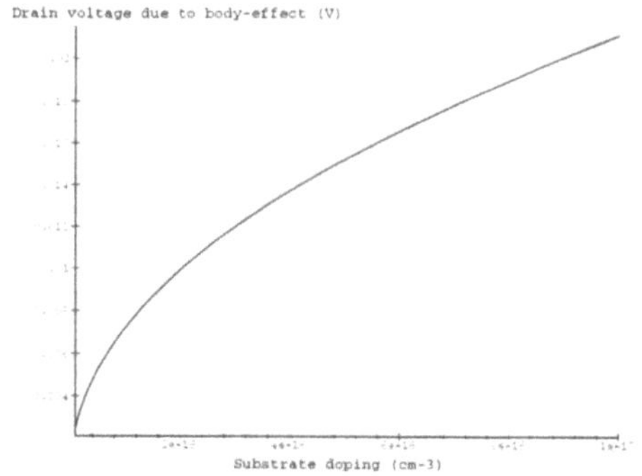

Figure 2.3

Noise at the drain due to body effect produced by V_{bs}=10 mV, as a function of the substrate doping.

2.1.2.c. Noise reception through contacts.

If supply lines powering the analog circuitry are also used to bias the substrate, noise may be acquired through the contacts and affect these lines and all the circuitry connected to them. If the affected circuit has a low PSRR, this may become the main noise reception mechanism. An accurate design with differential circuitry may eliminate the problem.

2.2 Coupling between MOS devices

2.2.1. General characteristics

To start substrate coupling characterization, data from device simulations is obtained. A structure consisting of a vertical cross-section of an IC with two close transistors is defined, as shown in Figure 2.4. The dimensions and technological characteristics are those of a 1.5 μm technology, and are detailed in Table 2.I.

The device simulator generates a grid over the cross-section defined, and solves Poisson's and carrier continuity equations on all the mesh nodes. The memory and CPU time demands of the simulations increase along with the number of nodes. Thus, in order to minimize the complexity of the structure simulated, we avoid substrate contacts and assume the substrate biased with a backside contact at the bottom of the device, as can be seen in Figure 2.4. For the same reason there has to be a substrate thickness of only 50 μm, much lower than usual. The substrate is always P-type,

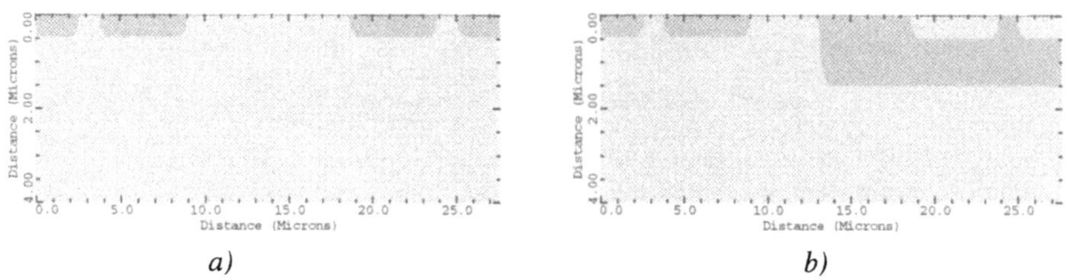

Figure 2.4

Device consisting of two transistors on the same substrate, to be analyzed with a two-dimensional device simulator. a) Two NMOS transistors. b) One NMOS and a PMOS inside an N-well

Table 2.1.

Transistor characteristics in the simulations performed in this section.

Gate length (L)	1.6 μm
Gate width (W)	2 μm
Gate oxide (t_{ox})	250 Å
Substrate doping (N_{sub})	1.5e16 cm^{-3}
N-well doping (N_{well})	2e16 cm^{-3}
Junction depth (x_j)	0.4 μm
Well depth (x_{jwell})	1.5 μm
Drain width (a)	4.4 μm
Distance between devices (d)	10 μm

except when simulating coupling between two PMOS transistors. Lightly doped or P- wafers, and heavily doped or P+ wafers are considered in the analysis. In the second case, substrate doping is about 10^{19} cm^{-3}, and the epitaxial layer is 4 μm thick, with a $1.5 \cdot 10^{16}$ cm^{-3} impurity concentration. P+ wafers are usually manufactured with epi layer thicknesses of about 10 μm, but the wafer processing considerably reduces the effective thickness, due to vertical diffusion effects.

The two transistors defined on a same bulk are connected to a SPICE-defined circuit, so one of them switches and generates disturbances, while the other one has a constant drain voltage. The switching transistor belongs to an inverter driven by a square signal. The sensitive device is saturated, and biased in such a way that the steady drain voltage is 2.5 V. Figure 2.5 shows a schematic diagram of this biasing net. In order to make the analysis simple and independent of package effects, contacts and power-supplies are connected to ideal voltage sources.

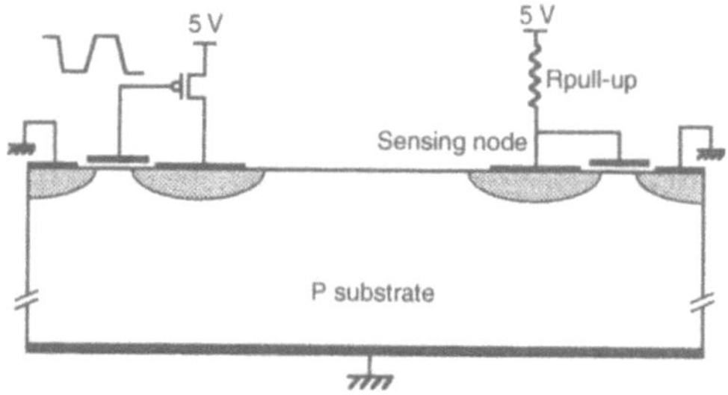

Figure 2.5

Schematic diagram of the circuit simulated, in which the NMOS transistors are defined with the device simulator and share the same substrate.

Figure 2.6 depicts the waveforms that appear at the different terminals. First the voltage at the input and output of the noisy inverter is shown in Figure 2.6.a. Next, in Fig. 2.6.b, the noise at the substrate is shown, the continuous line representing the noise under the disturbing NMOS, the dashed line representing the noise under the disturbed device. Observe that both waveforms are coincident, except that the second one has been attenuated in its flow along the substrate. The capacitive coupling mechanism may be verified by relating the voltage peaks in Figure 2.6.b to the transitions of the noisy drain voltage. Last, Figure 2.6.c shows the waveform at the drain of the saturated device, and we can see that, despite being circuit isolated from the inverter, disturbances are coupled for every transition.

The waveform observed at the sensitive device drain can also be interpreted in view of the noise reception mechanisms discussed in 2.1.2. Focusing on the noise produced by the first input transition, the two noise peaks at the substrate produce the appearance of three noise peaks at the sensitive drain, corresponding to each of the voltage transitions of the substrate voltage. Moreover, two voltage peaks are superimposed due to body effect: the small positive peak at the substrate causes a light voltage decrease at the drain, while the second negative peak produces a large voltage increment. This way the three noise peaks appearing in Figure 2.6.c for every transition are identified. The two first ones are mainly due to junction capacitive coupling, while the third one is mainly due to body effect.

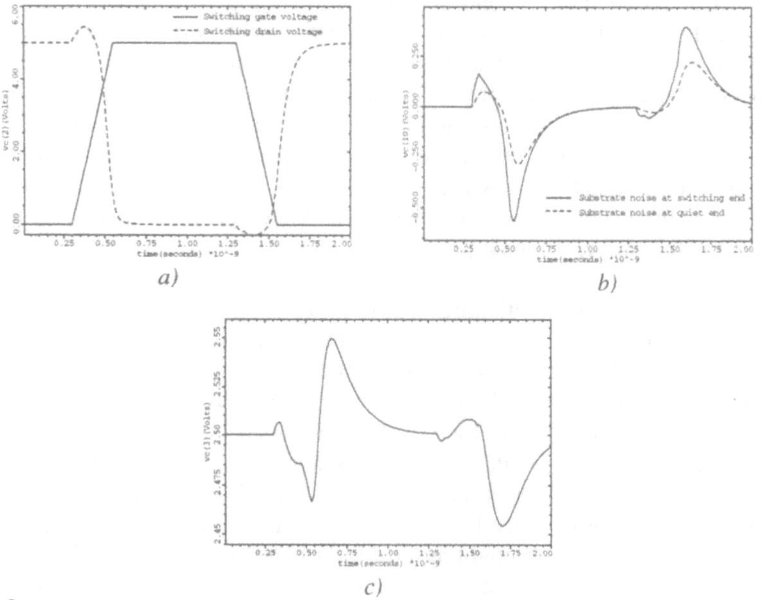

Figure 2.6

a) Waveforms at the input and output of the noisy inverter. b) Noise at the substrate. c) Noise received at the drain of the neighboring device.

2.2.2. Noise dependence on layout and technology parameters

In order to characterize noise coupling in MOS technologies, its dependence on a number of process or layout parameters is examined. We used both a device simulator and experimental measurements to perform this analysis. In the device simulations, we take a reference situation consisting of two coupled devices like those shown in Figure 2.4, with the characteristics detailed in Table 2.I. Starting from this reference case, a number of simulations modifying the value of each of the parameters studied are performed, while the remaining values are kept invariant. Curves showing the noise dependence on each of the parameters studied are obtained. The circuit used is the same as the one shown in Figure 2.5, except that the sensitive transistor working in its linear region, with a 0.2 V continuous drain voltage in an NMOS, and 4.67 V in a PMOS. The substrate is always biased by means of a backside contact, neglecting the package parasitics in order to simplify the problem as far as possible and thus make it independent of related effects.

The measurements are made on simple test structures similar to those simulated, consisting simply of pairs of coupled devices. Two circuits are manufactured, one with a 1.5 µm CMOS technology, and another with a 0.5 µm technology. Each of these contains different pairs of transistors or inverters, with different characteristics and placing. None of the chips is packaged, so we avoid any parasitic effects connected with the package and the output pad circuitry. This will mean always biasing the substrate with surface contacts, unlike the case in the simulations. Biasing and switching signal injection are made through static probe leads, while noise is measured with a dynamic low-capacitance probe. White noise is eliminated by averaging a high number of acquisitions, while probe-to-probe coupling was eliminated with a waveform post-processing.

2.2.2.a. Dependence on the distance between devices.

First the noise dependence on the distance between coupled devices is determined, with a length ranging from a few microns –the minimum distance allowed by design rules–, to a hundred microns. Figure 2.7 shows the results obtained in P- substrates, and for the different type of transistors. The noise coupled quickly drops with distance, not linearly, but with a stronger dependency, $1/(d^{1.5})$ approximately. This means that in this type of substrate, a first and effective measure for reducing noise would be to maximize distance between noise source and sensitive device. This may be enough to fully eliminate coupling problems if noise is introduced by the devices. Nevertheless, if noise is injected by substrate contacts, increasing the distance may be useless if contacts are still placed close to the sensitive devices.

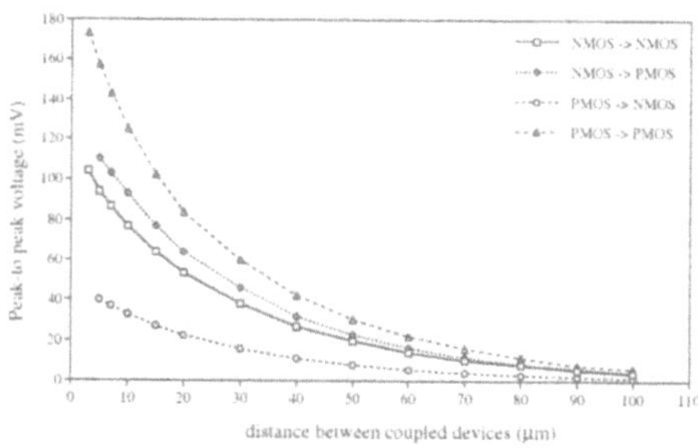

Figure 2.7

Substrate coupling dependence on the distance between devices, lightly doped wafer.

In heavily doped substrates, absolute noise level is drastically reduced due to the effect of the backside contact connected to the highly conductive bulk, which allows noise to be branched off to ground through a low impedance path. Figure 2.8 compares the dependence on the distance between two NMOS in both types of substrates, with the vertical axis in a logarithmic scale to better appreciate the trend. It can be seen that in very short distances, noise decreases faster in P+ substrates than in P- ones. This is because in this range noise is propagated along the epi layer, while

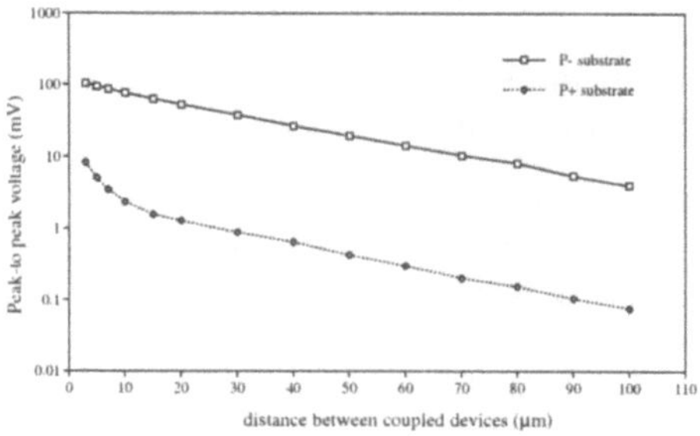

Figure 2.8

Substrate coupling dependence on the distance between devices. Comparison between P- and P+ substrates, in both cases biased by means of a backside contact.

when separation between devices is increased the fraction of the disturbance that penetrates the P+ bulk increases, thus reducing the noise received by the neighboring device. For distances larger than about twice the epi layer thickness –here we agree with [12]–, the dependence on distance is equal to what is found in P- wafers. Some authors apparently disagree when they claim that in epi wafers noise is independent of distance [13], [14]. This is true if noise does not find return paths to ground in its flow along the substrate –i.e. no backplane, or contacts presenting low impedance to ground. In the simulated situation there is an ideally grounded backside contact, and noise is thus attenuated in its propagation.

We will now present the measurement results on a test circuit manufactured with a 1.5 μm digital CMOS process. This is an N-well process, with 1.6 μm minimum gate length, that uses lightly doped P wafers. Six pairs of transistors are implemented in the circuit, as can be seen in Figure 2.9.a. The pairs combine NMOS and PMOS

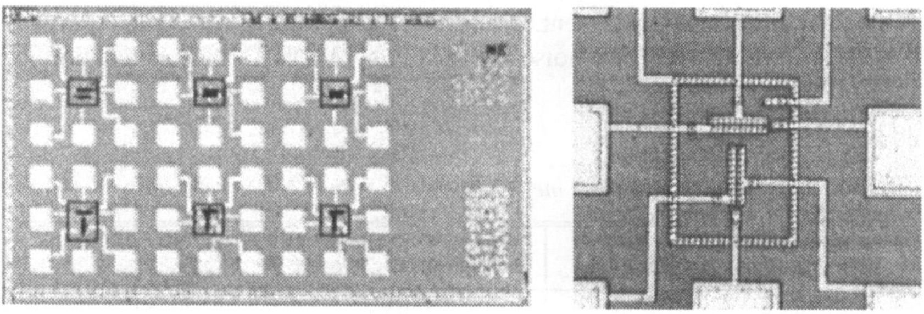

a) *b)*

Figure 2.9

Test chip manufactured in a 1.5 μm CMOS process. a) Global view. b) Detail of a pair of transistors.

transistors, and each combination is implemented twice, once placing the transistors with parallel gates, and once with transverse gates. The devices are 40 µm wide, and are separated by the minimum distance, 3.2 µm. Figure 2.9.b depicts a detail of two transistors in a perpendicular placement.

The noisy device is biased as an inverter, while the sensitive device is biased in its linear region. The inverter is driven by a clock signal with 100 KHz frequency and 1 ns transition time. For every pair, the sensitive transistor is biased in two different ways, with the source and drain assignments shown in Figure 2.10.

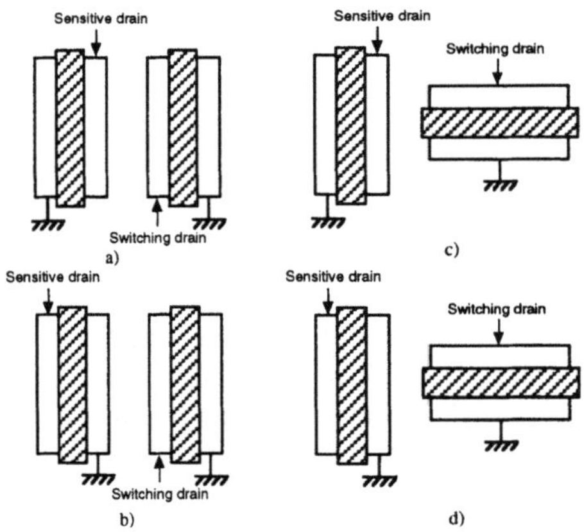

Figure 2.10

Pattern of the different ways in which coupled devices are biased.

Table 2.II summarizes the results obtained, while Figure 2.11 shows an example of the noise waveforms measured [15]. Noise levels are higher than those obtained from simulation, although they cannot be directly compared because the transistor sizes and substrate biasing are now different. As in the simulation results, noise between two PMOS is somewhat higher than noise between two PMOS. Device orientation is also

Table 2.II.

Results of substrate coupling measurements, 1.5 µm CMOS technology.

Transistor pair:	drain close to the noisy device	source close to the noisy device
parallel NMOS-NMOS	288 mV_{pp}	71.5 mV_{pp}
transverse NMOS-NMOS	51.5 mV_{pp}	31.5 mV_{pp}
parallel PMOS-PMOS	310 mV_{pp}	72.5 mV_{pp}
transverse PMOS-PMOS	43 mV_{pp}	33.5 mV_{pp}

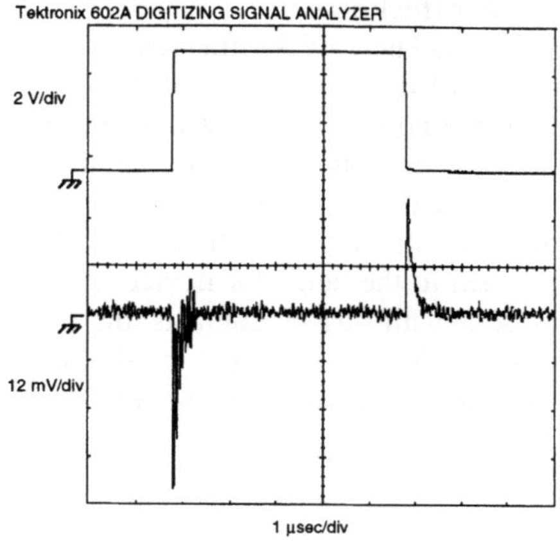

Figure 2.11

Disturbing inverter input voltage (up), and noise measured at the disturbed transistor (down).

seen to be of importance. It must be taken into account that, for parallel devices, the distance is very small, and thus any separation increase, like the one forced when altering device orientation, cuts down noise considerably.

The difference between the noise received when swapping the drain and source terminals over in the sensitive transistor is particularly significant. When placing the source between coupled drains, the disturbance path along the surface is obstructed and noise is forced to the substrate inside. This result can also be obtained from simulations. Table 2.III shows the results of four simulations for two coupled transistors, with the characteristics of the reference case in section, and for the four possible assignments that may be given to the four implants forming the transistors. It can be seen that the worst case is where drains are in front of each other, while the best involves inserting both sources between the noisy and sensitive nodes.

Table 2.III.

Simulation results showing the importance of source and drain placement.

noisy NMOS --- sensitive NMOS	
source - drain --- drain - source	77 mV_{pp}
source - drain --- source - drain	35.5 mV_{pp}
drain - source --- drain - source	34 mV_{pp}
drain - source --- source - drain	16.2 mV_{pp}

2.2.2.b. Dependence on wafer thickness.

As shown above, a substrate thickness of 50 μm was assumed in all simulations, this being a lower value than that of real wafers. Usual thicknesses are between 300 and 600 μm. Figure 2.12 shows the noise level for wafer thicknesses up to 500 μm [15]. The tendency observed towards noise increase is due to the backside biasing, because the deeper the backside contact, the worse the biasing at the surface, and the greater the noise. Above a certain limit, the resistance to the backside contact is so large that all the disturbance is branched to the neighbor device, thus the cause of the increase saturation. If biasing was done with surface contacts, the influence of wafer thickness could be neglected. In a conclusion, it is not advisable to bias low conductive substrates only with a backside contact, because disturbances are then branched off to neighboring devices.

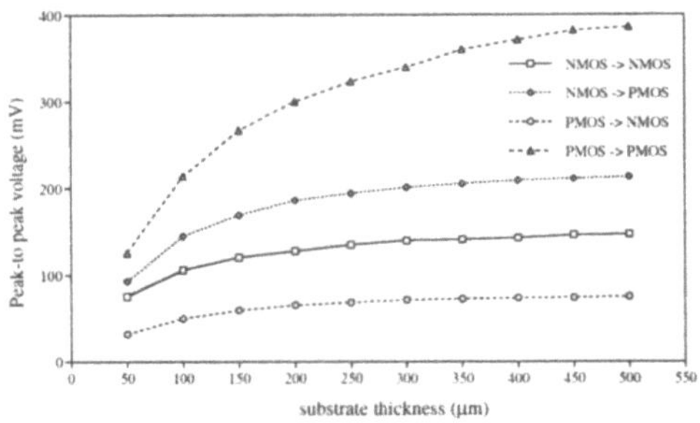

Figure 2.12

Substrate coupling dependence as a function of the thickness of a lightly doped wafer biased by means of a backside contact.

2.2.2.c. Dependence on the bias point of the sensitive device.

A number of simulations of the reference situation are performed, but now varying the sensitive device gate voltage between 0 and 5 V. Figure 2.13 shows the results obtained, while Figure 2.14 depicts the transfer curves of both devices. It can be observed that, both in NMOS and in PMOS transistors, noise is lowest when the device is in its linear region, while it is highest in the saturation region. In the NMOS device the three regimes are more clearly delimited, as the noise increase in the saturation region is higher. This can be explained by relating the noise increase to device gain, which is greater in an NMOS than in a PMOS, and which is coincident with the saturation region of each device. As a general conclusion, a device will have maximum substrate noise sensitivity when working in its saturation region, and the larger the gain, the larger the sensitivity.

The former simulation results have been complemented with measurements on a test IC manufactured with a 0.5 μm CMOS process. It uses P+ epi wafers, and devices are designed to be powered at 3.3 V. The IC contains six test structures, each one

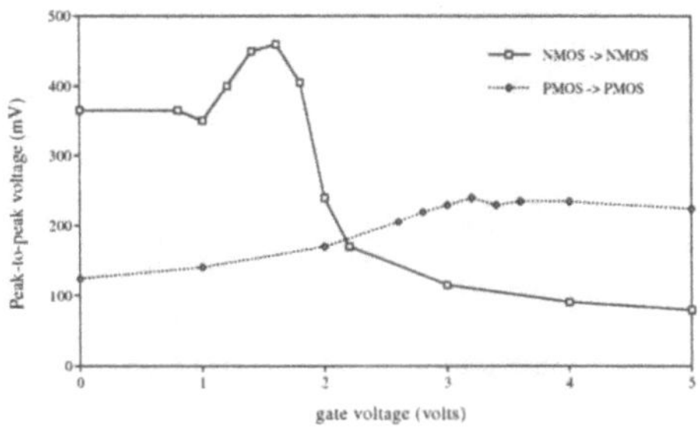

Figure 2.13

Substrate coupling dependence on bias point of the sensitive device.

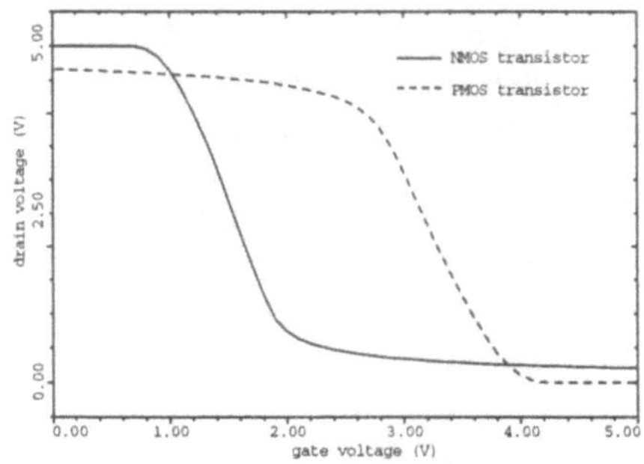

Figure 2.14

Transfer curves of the NMOS and PMOS transistors.

consisting of a CMOS inverter disturbing a sensitive transistor, every test structure being enclosed with the lead pads for biasing, noise injection and measurement. Figure 2.15 shows a view of part of the circuit, along with a detail of one of the structures.

The transistors in the noisy inverter have an aspect ratio of 10/0.5, and are driven by a clock signal with 1 MHz frequency and 1 ns rise/fall time. Sensing transistors have an adjustable gate voltage, so they can be biased at different points of their transfer characteristic. They are connected to a pull-up PMOS –or pull-down NMOS, depending on the case–, with lower gain and a fixed $|V_{gs}|$=3.3 V. Figure 2.16 is a diagram of the test circuits implemented.

Every test structure is different to the others in the transistor placement. Two of the structures have their sensor with the gate parallel to the transistors in the inverter, inside the same well. Another two structures have the sensor transistor perpendicular to the others, also inside the same well as the disturbing ones. Lastly, a further two

Figure 2.15

Test IC manufactured with a 0.5 μm CMOS process. a) Partial view. b) Detail of one of the test circuits.

structures have two N-wells, one for the noisy PMOS and the other for the sensor circuit, cross-placed as on a chess board. Two structures are implemented for each of these combinations, one with the sensor circuit in Figure 2.16.a, and the other one with the circuit in 2.16.b. Figure 2.17 shows the characteristic transfer curves of both sensing circuits. It can be seen that the gain of the structure in Figure 2.16.a is much lower than the gain of the circuit in 2.16.b, due to the lower PMOS transconductance.

Figure 2.16

Test circuit implemented. a) NMOS acting as a sensor. b) PMOS acting as a sensor.

Three measurements are made on every test structure: one biasing the sensing device with a 0 V gate voltage, another with a 3 V biasing, and a third one driving the sensor to its maximum gain point. Table 2.IV shows a summary of the results obtained. Each of the values shown is the average of at least three measurements.

If the results according to transistor placement are compared, we can see that placing the P transistors in different wells helps to reduce noise. On the contrary, parallel or transverse placement do not have any significant influence. This is because the P+ wafer makes the direction or distance between devices insignificant.

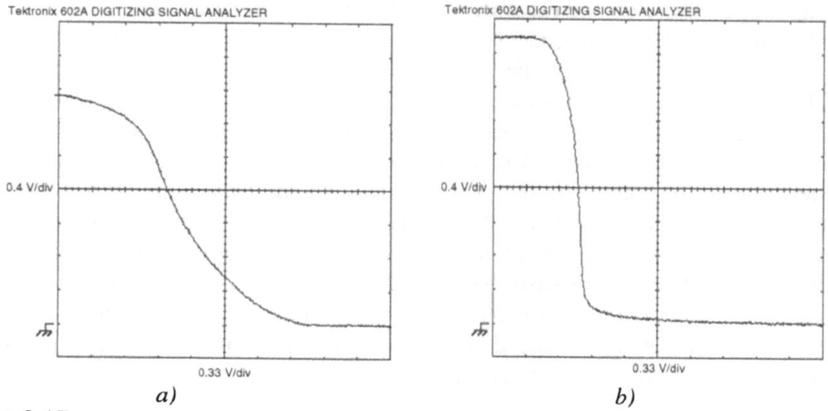

Figure 2.17

Static transfer characteristic, $V_{out}(V_g)$, measured at the sensor circuit in Figure 2.16.a (a), and in the sensor circuit in Figure 2.16.b (b).

Table 2.IV.

Summary of the measurements obtained on the test IC in a 0.5 μm CMOS technology.

Sensor transistor →	parallel PMOS	transverse PMOS	PMOS, two N-wells
V_g=0 V	171 mV_{pp}	195 mV_{pp}	175 mV_{pp}
V_g≈1 V	269 mV_{pp}	266 mV_{pp}	226 mV_{pp}
V_g=3 V	243 mV_{pp}	241 mV_{pp}	135 mV_{pp}
Sensor transistor →	parallel NMOS	transverse NMOS	NMOS, two N-wells
V_g=0 V	166 mV_{pp}	159 mV_{pp}	152 mV_{pp}
V_g≈0.9 V	607 mV_{pp}	635 mV_{pp}	583 mV_{pp}
V_g=3 V	206 mV_{pp}	196 mV_{pp}	183 mV_{pp}

As regards the type of transistor and the bias point, it is observed that the greater noise is always obtained at the maximum gain point in the transfer curve. Moreover, the greatest amount is received by NMOS transistors, because of their larger gain. Thus, the gain of the transistors at their bias point significantly determines the amount of noise received, as was concluded from the simulations. Pull-up or pull-down transistors do not affect the situation because they are biased with gate voltages of $|V_{gs}|$=3.3 V and then their bias point is placed in regions where the gain is very low or zero.

2.2.2.d. Dependence on the rise/fall time of the noisy signal.

Maintaining the same reference situation, simulations are performed in which the transition time at the gate of the noisy device is varied from 10 ps to 10 ns. The switching speed at the gate will influence the switching speed at the output drain, which is the terminal actually injecting noise. For the range of variation taken, the drain falling edge goes from 45 ps to 1.5 ns. The relation is not linear, as very low transient times are limited by the characteristics of the transistors. It is observed in Figure 2.18 that the coupled noise sharply increases with the transition speed, reaching hundreds of millivolts. Even when output drain speed is saturated, noise still increases, which means that part of the disturbance is directly introduced from the gate. Observe that the range of variation of noise level on the vertical axis is the largest seen so far, which means that switching speed is the most important parameter determining the amount of noise coupled through the substrate.

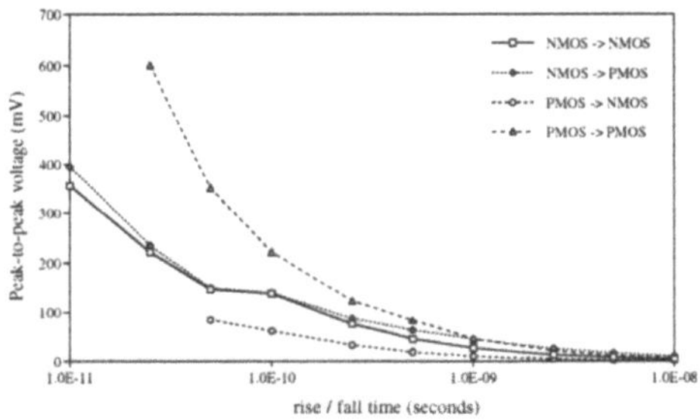

Figure 2.18

Substrate coupling dependence on the switching speed at the gate of input of the noisy inverter.

2.3 Coupling between devices in BiCMOS technologies

In this section it is studied the coupling that can be produced between bipolar and field-effect transistors in BiCMOS technologies. In bipolar transistors, noise is mainly injected and received through the junction capacitances. The results obtained in the previous section not connected with specific characteristics of MOS transistors will also be extensible to BiCMOS technologies. The same type of dependence on distance or transition time is thus expected.

A structure consisting of two coupled devices is taken, and transient analysis is performed to determine the dependence of noise on different parameters. The structure consists of an NPN vertical transistor sharing the silicon substrate with an NMOS transistor, both devices 40 μm apart. The dimensions and technological characteristics of both transistors are taken from a 3 μm BiCMOS process. The substrate used by this process is a heavily doped epitaxial one, and it is biased through a backside contact, with a series inductance of 10 nH to model the package parasitics in a first approximation. Figure 2.19 shows a diagram of the vertical cross-section of the bipolar transistor. We must stress the collector structure, formed by a lightly doped deep implant (NTUB), plus a highly conductive buried layer –$8.5 \cdot 10^{18}$ cm^{-3} doping– and a deep contact, also highly conductive (NPLUG). This way the collector resistance r_c is minimized. The total device length is 55 μm.

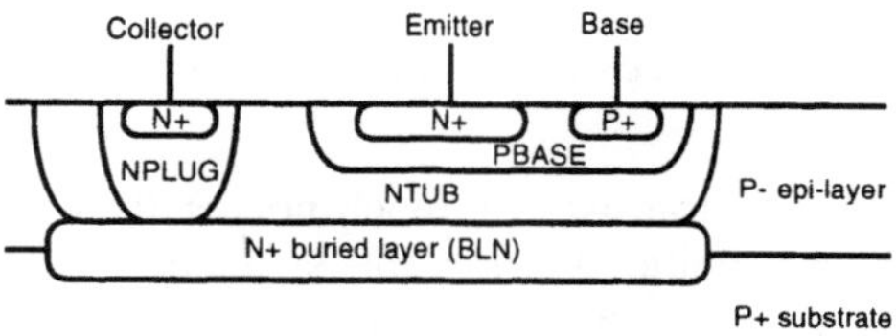

Figure 2.19

Diagram of the vertical cross-section of the NPN bipolar transistor used in the simulations.

The analysis is divided according to the two possible directions of the coupling, i.e. coupling from the bipolar to the MOS transistor, or the other way round. For every situation, we will consider the bipolar transistor common-emitter or common-collector biased.

2.3.1. Bipolar transistor as a noise source

2.3.1.a. Common-emitter biasing.

First the noise produced by a BJT when the collector is switching is analyzed, in a simple common-emitter configuration like the one shown in Figure 2.20. A low-high transient is forced at the base, with a rise time of 1.5 ns typical of this technology. This transition drives the transistor from cutoff state to saturation, producing a high-to-low transient at the collector. The noise waveforms obtained show a maximum peak of 30 mV (55 mV_{pp}). In order to compare this with the noise that a MOS transistor would produce in these circumstances, a simulation to measure coupling between two NMOS in this technology is performed, with a transition time at the gate of 0.5 ns, devices 40 μm apart, and the same type of substrate biasing. The amplitude of the noise peak measured is 2.6 mV, i.e. the noise produced by a switching bipolar transistor is at least an order of magnitude higher.

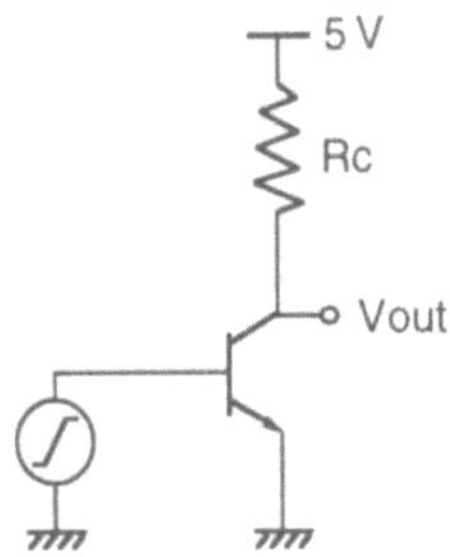

Figure 2.20

NPN bipolar transistor in a common-emitter biasing.

The reason why the bipolar transistor is much noisier than the MOS one is its size. Given that in a common-emitter biasing the switching collector forms a junction directly with the substrate, then the greater the junction area, the greater the amount noise injected. To confirm the importance of the collector characteristics, we analyze the noise dependence on the buried layer doping, and on the presence of a deep collector contact (NPLUG in Figure 2.19). Figure 2.21 shows that noise increases with the buried layer doping as well as with the presence of the collector contact [16]. The deep contact helps to reduce the collector resistance r_c and thus increase the speed at this node, which increases coupling through the junction. The same thing occurs with the buried layer doping, which not only reduces the collector resistance, but also the junction capacitance.

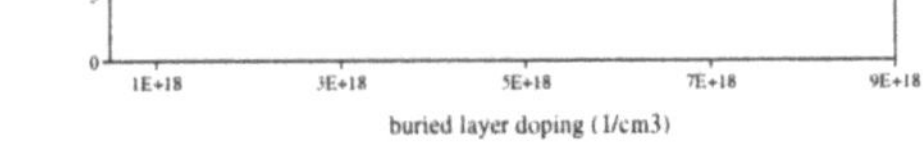

Figure 2.21

Coupling dependence on buried layer doping, common-emitter biasing.

2.3.1.b. Common-collector biasing.

A similar analysis is now performed, but with the BJT in a common-collector configuration, as shown in Figure 2.22. The exciting signal is the same as before, and now the emitter is the output (switching) node.

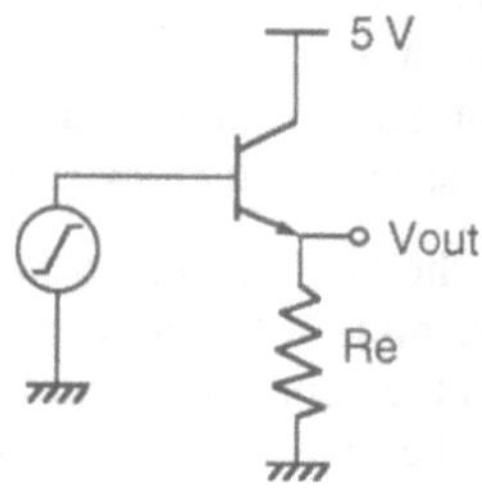

Figure 2.22

NPN bipolar transistor in a common-emitter biasing.

The results obtained are represented in Figure 2.23, again as a function of the buried layer doping and of the presence of a deep collector contact [16]. Firstly, a significant noise level reduction is observed. This is because now the collector voltage is fixed, and only the internal part of the transistor switches, but not the part in contact with the substrate. As regards the dependence on r_c, the tendency is now opposed to what is observed in common-emitter biasing. The noise generated increases when the doping of the buried layer and when the deep contact NPLUG is removed.

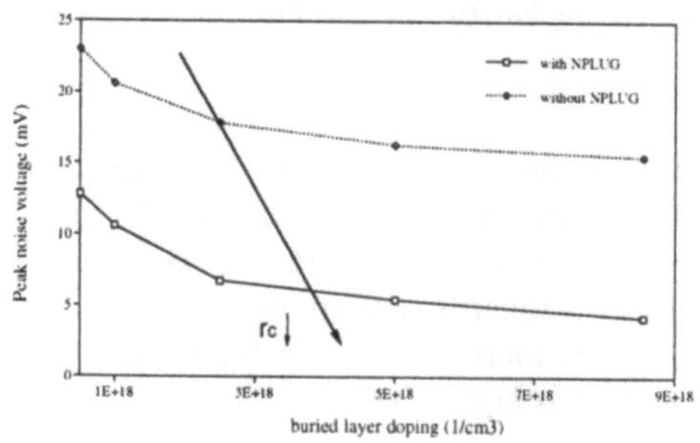

Figure 2.23

Coupling dependence on buried layer doping, common-collector biasing.

2.3.2. The MOS transistor as a noise source

After determining that a bipolar transistor is indeed capable of coupling noise to the substrate, we shall now look at its sensitivity to noise. Taking the same structure as before, simulations are performed when the NMOS transistor switches, with a gate transition time of 0.5 ns. The bipolar transistor is biased in its linear region with a 1 V base voltage.

In a common-emitter biasing, the noise amplitude received at the collector is 5.75 mV, i.e. it is greater than the noise between two NMOS. The result is reasonable as noise couples through a capacitance larger than that of an MOS transistor drain. There is a slight tendency for noise to increase when increasing the doping.

In a common-collector biasing, the noise received is much smaller, around 1.65 mV. This is because now the sensitive node is the emitter, which is located in the inner part of the device. Figure 2.24 shows the dependence on buried layer doping. As expected, increasing the doping decreases the resistance to the voltage source, and consequently reduces the amount of noise reaching the emitter.

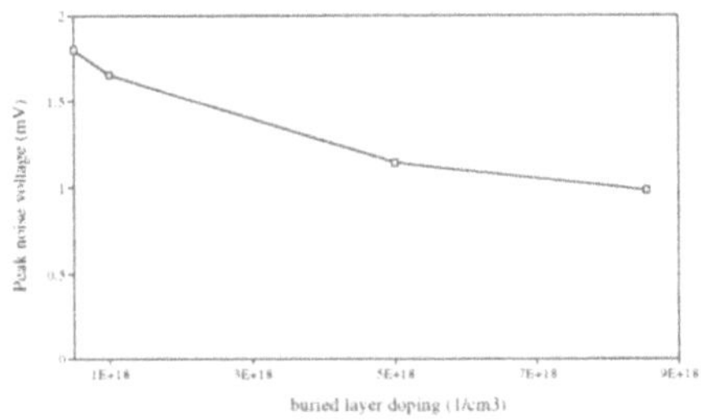

Figure 2.24

Coupling dependence on buried layer doping in the sensitive device, in a common-collector biasing.

As a summary, Figure 2.25 shows the noise level comparison between the different types of transistors and the different biasing. A bipolar transistor is noisier than a (minimum size) MOS one, especially when the collector is the switching node (common-emitter biasing). As regards sensitivity to noise, given the difference of sizes the bipolar collector receives more noise than the MOS drain. On the contrary, the noise immunity of the emitter is higher because the collector encloses and shields it.

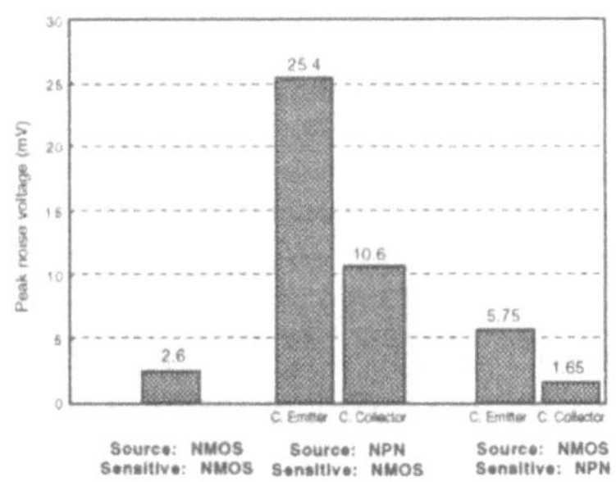

Figure 2.25

Coupled noise comparison between the different situations considered in this section.

2.3.3. Measurements in a test circuit

To cross-check simulation results, another test IC was manufactured, this time in a 3 μm BiCMOS process. The wafer type is P+, and circuits are driven at 5 V. Once again, the chip consists of a number of disturbing circuits, in this case BiCMOS inverters, near sensitive devices, in this case NMOS transistors with a 100 KΩ pull-up resistor implemented with a PMOS. Figure 2.26 shows a diagram of the BiCMOS inverter topology chosen, together with the sensor circuit. One should observe that the inverter contains a common-collector biased bipolar transistor (*Q1*), and another common-emitter biased one (*Q2*). The first will conduce when a low logic level is applied at the input, while the second one goes from cutoff to saturation in a rise transient of V_{in}. Voltage supply, signal injection and noise sensing are again performed with probe leads. The exciting signal is a clock with a 200 KHz frequency and 50 ns transient time. Switching times at the output are considerably lower, between 11 and 22 ns, depending on the transistors and transition direction.

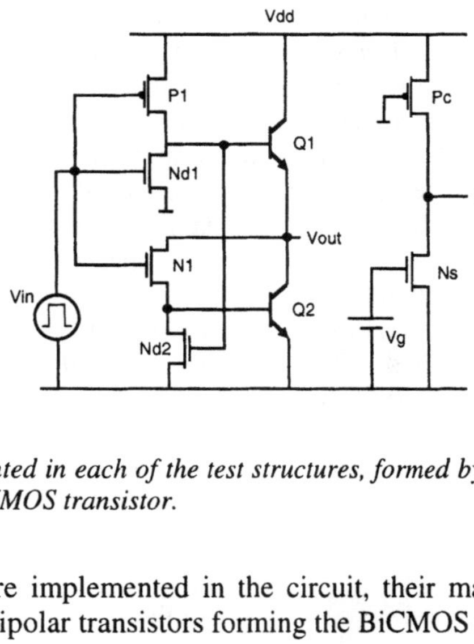

Figure 2.26

Circuit implemented in each of the test structures, formed by a noisy BiCMOS inverter and a sensitive NMOS transistor.

Four test structures are implemented in the circuit, their main difference being the characteristics of the bipolar transistors forming the BiCMOS inverter.

- Pair A: Bipolar transistors have a minimum emitter area (7 μm × 7 μm) and deep collector contact (NPLUG), so their collector resistance is r_c=1 KΩ.
- Pair B: Like pair A, except that NPN transistors have shallow collector contacts, so the collector resistance is now r_c=1.8 KΩ.
- Pair C: Like pair A, but now the emitter area is ten times larger.
- Pair D: Like pair A, but now the sensitive device is enclosed by a P+ guard ring, to observe its noise reducing effect.

Figure 2.27 shows two details of the circuit, where pairs A and C may be appreciated.

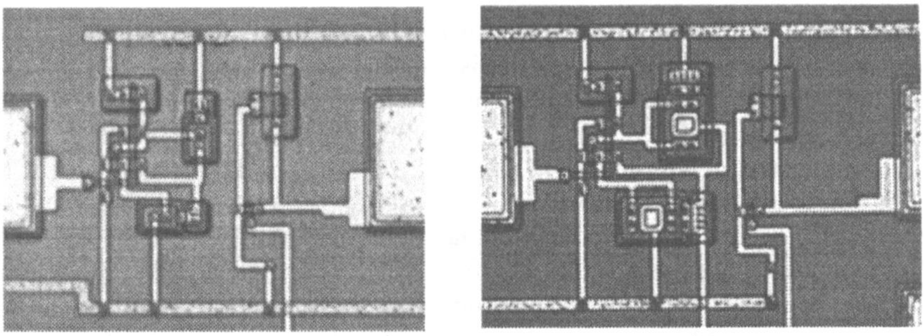

a) b)

Figure 2.27

Details of the circuit manufactured in 3 μm BiCMOS technology. a) Pair A. b) Pair C.

First, the noise measured in each of the four structures is compared, with the results shown in Figure 2.28 [16]. It can be seen that transistors in pair C, which are the largest ones, are the noisiest. The simulation analysis had already shown that the collector-substrate capacitance, and thus the transistor size, play a major role in the amount of noise injected. Concerning the importance of collector resistance, its importance can be determined by observing the effect of the deep collector contact. In 2.3.1 we had determined that its absence mainly influenced the common-collector, increasing noise level. This is proved in the measurements, because the noise produced by the inverter without deep contacts (pair B) is greater than the noise when the contact is present (pair A). Lastly, we can see that a grounded P ring provides a noise reduction of about 30%. This relatively low reduction is typical of heavily doped substrates, as will be seen below.

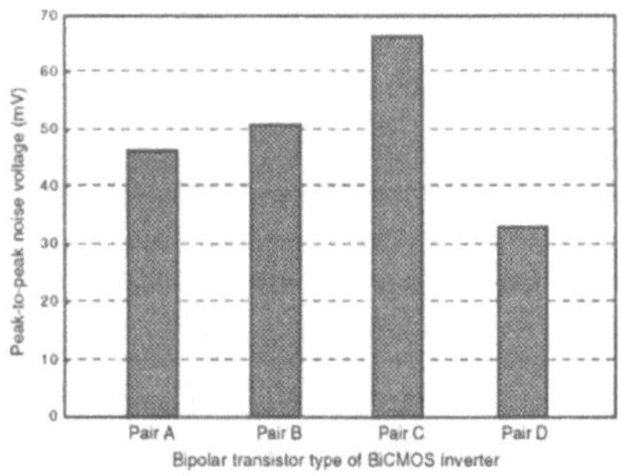

Figure 2.28

Comparison of noise levels obtained in the four test structures.

We will finally present two other results that can be connected with those obtained in section 2.2. Firstly the noise dependence on rise/fall time of the signal at the inverter input is obtained. As could be expected, noise decreases when transitions are slowed down, in all three types of transistors. Figure 2.29 shows the results obtained for pair C, although the dependence obtained for the other transistors is similar. Secondly, and taking advantage of the fact that a sensitive transistor with the same characteristics is located in every test structure, we find the noise dependence with the distance between the noisy inverter in pair C, and the other noise receptors. The results are shown in Figure 2.30, and the noise can be seen to decrease slightly with distance, which is typical of heavily doped substrates.

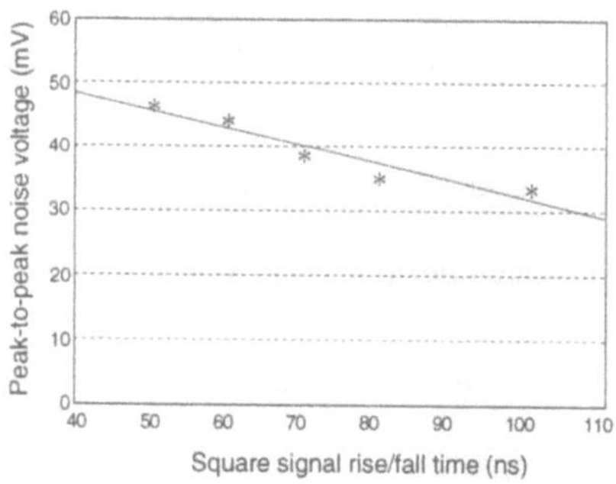

Figure 2.29

Coupled noise dependence on switching speed at the input of the noisy inverter.

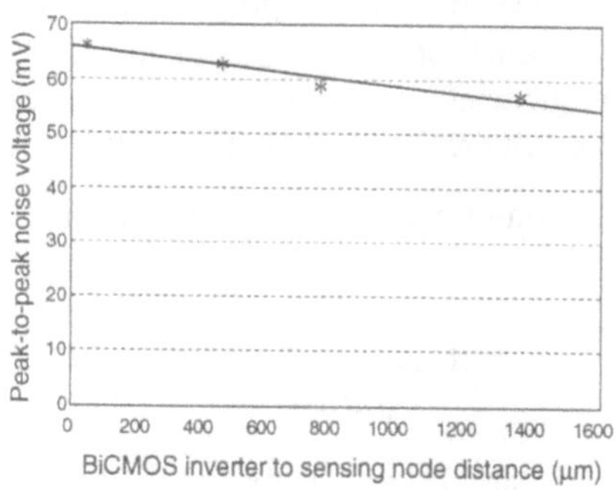

Figure 2.30

Coupled noise dependence on the distance between devices.

2.4 Noise propagation analysis

2.4.1. Some concepts of noise conduction in silicon

Like other conductors, silicon displays three different behavior regimes depending on the frequency of the electric field applied [17]:

- slow-wave regime
- skin-effect regime
- dielectric regime

Each of these regimes will now be analyzed. The current density due to a number of carriers N_e is

$$J = N_e q v = \sigma E \qquad \text{Eq. (2.5)}$$

where the conductivity σ has been defined as [18]

$$\sigma = \frac{N_e q^2 \tau_m}{m_e} \qquad \text{Eq. (2.6)}$$

If the term due to bound carriers is added (displacement current), the well known distributed form of Ohm's law is obtained,

$$J = (\sigma + j\omega\varepsilon)E \qquad \text{Eq. (2.7)}$$

This means that a fragment of silicon may be electrically modeled as a conductance and a capacitance. Nevertheless, for usual doping levels, conductivity assumes a value around 10 $\Omega^{-1}m^{-1}$, while product $\omega\varepsilon$ has a value of just 0.65 $\Omega^{-1}m^{-1}$at a frequency of 1 GHz. In the range of frequencies usually encountered in ICs, the electrical behavior of the substrate is therefore fully resistive [14]. This is the slow-wave regime.

As frequency increases, skin-effects appear. This means that electric fields are restricted to a zone of depth

$$\delta = \frac{1}{\sqrt{\pi f \mu \sigma}} \qquad \text{Eq. (2.8)}$$

where μ is the material permeability. For depths up to δ, the behavior continues to be resistive, while above this value the behavior is insulating. This is the skin-effect regime. Observe that the depth δ decreases with frequency, so only the highest frequency components in a noise pulse will be affected by this effect. The border between slow-wave regime and skin-effect behavior will be given by the frequency at which δ equals the wafer thickness. If a conductivity of 10 $\Omega^{-1}m^{-1}$ and a thickness of

500 μm are taken, the transition between both regimes is produced at f=100 GHz. If the substrate is a heavily-doped one, the transition is produced at much lower frequencies, around 1 GHz. In any event, the transition between both regimes is not sudden but gradual, and it is difficult to fix an exact threshold. If the substrate is biased through a backplane contact, in the skin-effect regime the disturbance cannot reach the chip backside, instead being branched off along the circuit surface to other neighboring devices or contacts, along a path of decreasing depth as frequency increases. Furthermore, given the frequency, noise can be transmitted as in a transmission line, which would reduce its propagation speed.

Two significant conclusions can be drawn from this discussion. Firstly, the effect of backside contacts is practically eliminated at high enough frequencies, resulting in worse substrate biasing. Secondly, in heavily doped substrates the frequency at which this regime is initiated is low enough to consider that the beneficial effects of a backside contact may be diminished or even eliminated in practice, depending on the characteristics of the wafer.

At high enough frequencies, the real and imaginary parts in equation (2.7) are made comparable. This is the dielectric regime, because silicon stops behaving like a distributed resistance and starts behaving like a capacitance. If σ=10 $\Omega^{-1}m^{-1}$, the frequency at which both terms on the right-hand side of equation (2.7) are made equal (relaxation frequency) is 15 GHz. In heavily doped substrates the relaxation frequency is much higher and this regime is never reached in practice.

2.4.2. *Three-dimensional analysis of substrate noise propagation*

2.4.2.a. Characteristics of the analysis

In this section an analysis of noise propagation is presented, leaving aside its injection and reception by devices. We will determine the influence of the wafer type, biasing, and characteristics of ground contacts on noise propagation. Given a simple layout, and an amount of noise at a given point, we wish to establish how noise is distributed along the silicon surface. We assume that noise is attenuated along the substrate the same way as a static potential. This quasi-static approximation will be valid whenever the substrate behaves resistively, i.e. in the slow-wave regime according to the classification seen above. This assumption does not imply a significant limitation as this mode applies for most of the noise frequency components.

The analysis used in this section is performed with the RAPHAEL program [19]. This tool uses the finite difference method to solve Poisson's equation in a volume. It does not solve carrier continuity equations, meaning that it does not allow semiconductor materials to be considered, but this is not important as substrates can be modeled as conductors with a certain resistivity. Thus, no distinction between N and P type substrates will be made.

As a methodology, we will take a reference situation, and alter geometrical parameters to observe their effects. The situation taken as a reference is a silicon fragment 300 μm long (x dimension), 200 μm wide (y dimension), and 50 μm thick (z dimension). A small electrode is placed at one of the ends (x=0 μm), biased to 1 V. This electrode will stand for a noisy node, and its voltage will be taken as a reference of the amount of noise injected, which is attenuated along the substrate. Thus, if a 0.1 V voltage is measured at a point on the silicon surface, this means that the noise at that point is 0.1 times the noise injected. At the opposite edge of the fragment (x=300 μm) we fitted a grounded contact, which can be interpreted as a substrate contact at the chip periphery. In the majority of the situations analyzed there is a grounded guard ring, consisting of an electrode of variable length and centered in the y axis, as shown in Figure 2.31.

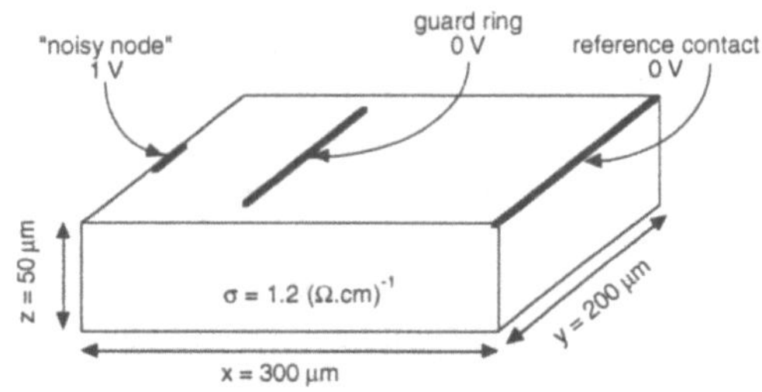

Figure 2.31

Schematic diagram of the silicon fragment simulated, taken as a reference case.

Figure 2.32 shows the constant voltage contour lines at the surface after simulating a situation like that of Figure 2.31. It can be seen that voltage drop concentrates around the noisy node, and that noise levels are low in the area between the guard ring and the end of the fragment (x=300 μm), thus being protected from the disturbance.

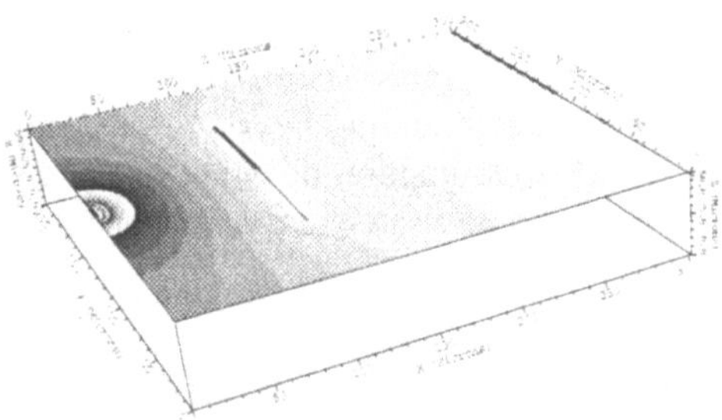

Figure 2.32

Example of the RAPHAEL simulation of the potential drop along the surface of a silicon fragment like the one drawn in Figure 2.31.

2.4.2.b. Ground contacts and noise propagation

In section 2.2 we verified that the presence of backside contacts makes noise attenuate with distance in P+ and in P- substrates. When no backside contact is present, noise seems to attenuate with distance in P- substrates, while it remains constant in P+ epi wafers [12]. It is particularly important to stress this point to understand the factors determining noise attenuation. The fact that in P- substrates the noise decreases with distance is not so much due to its high resistivity, but to the uniformity of the resistivity, together with the biasing characteristics. Assuming a silicon fragment like the one in Figure 2.32 with *only* the noisy contact at 1 V, noise would not be attenuated, but would be 1 V at any point in the fragment. In our case substrate is biased at the opposite end of the fragment (x=300 μm), so noise is progressively attenuated with distance, as shown in the vertical slice in Figure 2.33.a. If the disturbance encounters other contacts or guard rings, it is more quickly attenuated because these new contacts provide the current a return path of smaller impedance than the contact at x=300 μm. This is illustrated in Figure 2.33.b. The attenuation will be stronger the greater the number of contacts. Similarly, a backside contact forces faster noise attenuation, and its effect is greater the thinner the wafer, because then the impedance of the return path is smaller. In an infinitely thick wafer noise attenuation with distance would be independent of the presence of a backplane, while in a zero thickness wafer, noise drop would have an infinite gradient. Given the resistivity of P-substrates and the usual wafer thicknesses, we are actually closer to the first situation than to the second one. Noise attenuation with distance is not only important because it can be used as a protection measure –maximizing the distance between noisy and sensitive parts–, but also because it favors differential noise.

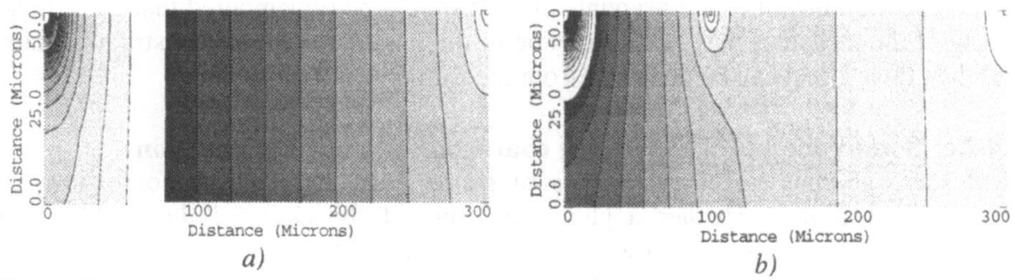

Figure 2.33

Constant voltage contour lines in a vertical slice of the P- silicon fragment, where the progressive attenuation of noise is appreciated, without the presence of any contact in the propagation path (a), and with the presence of a contact (b).

In a P+ wafer, and going back to the situation in which there are only the noisy node at x=0 μm and the reference contact at x=300 μm, the noise injected crosses the epi layer and is propagated along the low resistive bulk. This way attenuation is not uniform, but higher in the highest resistivity zones, that is to say, when crossing the epi layer. In the heavily doped bulk noise is also attenuated, but given the difference between the resistivities, it can be considered practically constant with distance.

The presence of other contacts or rings on the surface does not have the same effect as in a P- wafer, because now the impedance of the return path is independent of the location and only depends on the contact dimensions. Increasing the number of contacts will mean a lower substrate noise, although the reduction achieved will depend on the relative increase.

In a P+ wafer with a grounded backside contact, noise reaching the heavily doped bulk does find a return path whose impedance increases slightly with distance. If this path is better than the one offered by surface contacts, adding contacts (or guard rings) should not reduce substrate noise any further. In any event this depends on the impedance of this new return path through the backside contact. Compared to P- substrates, the lower resistivity in P+ wafers makes the effect of the backside contact significant despite its thickness. Nevertheless, the parasitic effects of the connection between the contact and external ground can make the backplane useless. These parasitic elements may increase the impedance of the backside return path to end up in a situation equivalent to absence of backplane. In the same way, in P- wafers the noise attenuation produced by surface contacts will depend on the parasitics of the respective connections. In the extreme case in which the impedances are infinite, noise does not attenuate with distance.

In short, the effects of noise reduction and attenuation will strongly depend on the impedances of the return paths offered to noise. In this chapter we have assumed ideal contacts, so only the effects connected with the technological characteristics of the noise reduction measures are accounted for. It should be remembered that the efficacy of any of the measures may be altered due to the impedance of the substrate contacts, and then their effect can be deducted from the former discussion.

2.4.2.c. Some concepts of guard ring characteristics and optimization

Next some geometrical characteristics of guard rings are studied. For example, a common question is whether a guard ring must fully surround the noisy or the sensitive part, or if a contact forming a barrier that collects noise is enough. Some simulations are made taking rings of different lengths, placed at a fixed distance to the noisy node (100 μm). The results obtained in a P- substrate are shown in Figure 2.34, while Figure 2.35 is of a heavily doped epi wafer, no backplane in both figures. The voltage on the surface along the axis y=100 μm is represented in all cases.

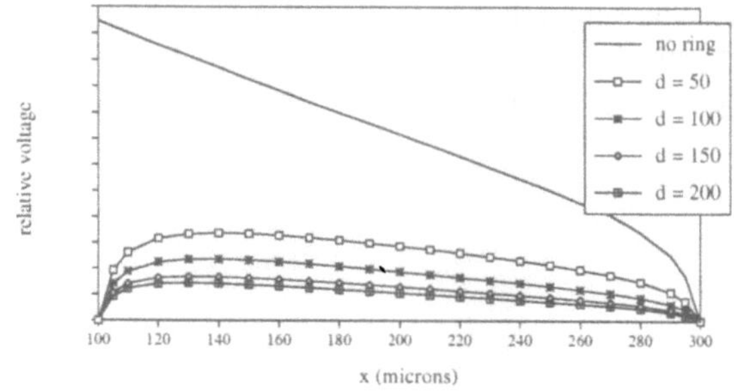

Figure 2.34

Effect of guard rings of increasing length, in a P- substrate without a backplane.

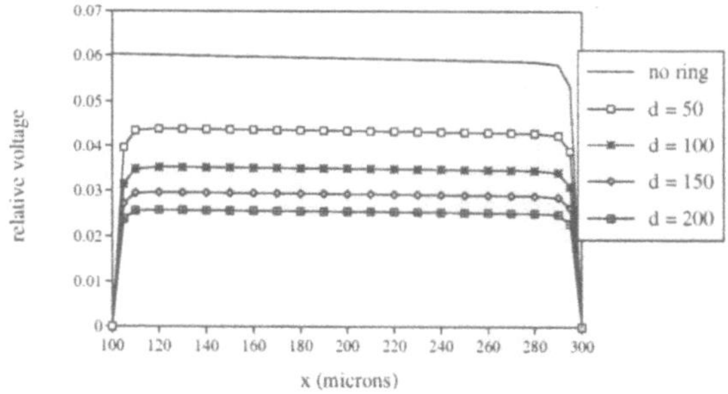

Figure 2.35

Effect of guard rings of increasing length, in a P+ substrate without a backplane.

It can be seen that the importance of the length of the guard rings depends on the situation. In all cases, the larger the ring, the lower the noise. Nevertheless, in P- substrates, rings above 50 μm add little noise reduction, while in P+ wafers the noise reduction is practically linear with the ring length. This confirms what was previously stated about the influence of contacts on noise propagation. In P+ substrates noise is not attenuated with distance, and its level depends on the total area of the contacts, regardless of their location –talking about *guard rings* will be meaningless–. On the contrary, if noise is attenuated with distance, as in P- wafers especially if they have a backplane, the disturbance is greatest on the straight line joining the noisy and the sensitive points, and the further the contacts from this line, the lower their effectiveness.

Another effect which is interesting to observe is the proximity of the guard ring to the noise source. Figures 2.36 and 2.37 show the noise at the surface for different distances between a 50 μm long ring and the noise source. In a uniform P- substrate, noise is lower the closer the ring to the noisy point, because then the impedance of the return path through the ring is lower. In P+ substrates the position of the ring shows again to be indifferent. In any event, the differences observed in Figure 2.36 are not great. This means that in practice the relative position of the ring is of little importance for the designer.

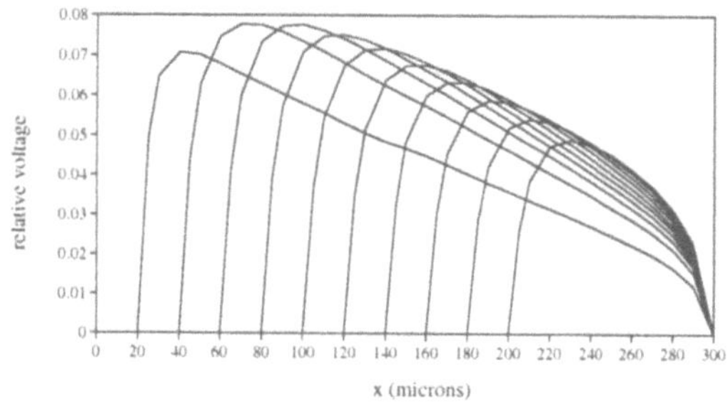

Figure 2.36

Effect of the proximity of the ring to the noisy point, in a P- substrate.

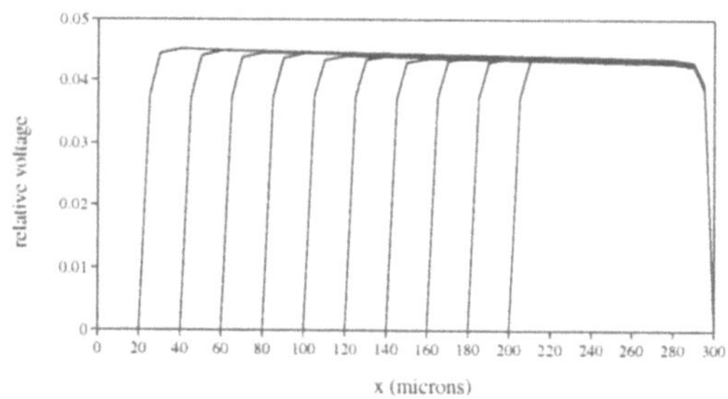

Figure 2.37

Effect of the proximity of the ring to the noisy point, in a P+ substrate.

It seems clear that the implementation of a guard ring is a good way of reducing substrate coupling, but implementing a large guard ring may lead to connectivity problems when generating the layout. In some cases it may be of interest to know how much effectiveness is lost because of a discontinuity in the ring, or when replacing the continuous ring by a row of contacts. In order to get an idea, a number of situations are simulated, starting from a situation with a lightly doped substrate without a backplane and with a guard ring 100 μm long. The results are compared with those of situations in which the ring is replaced by five contacts 12.5 μm apart, where replaced by three contacts 35 μm apart, or where replaced by only two contacts 80 μm apart. Every contact is 10 μm long, so the distance between the ends is always 100 μm. Figure 2.38 shows the constant voltage contour lines at the surface, while in Figure 2.39 the voltage at the surface along axis y=100 μm is represented.

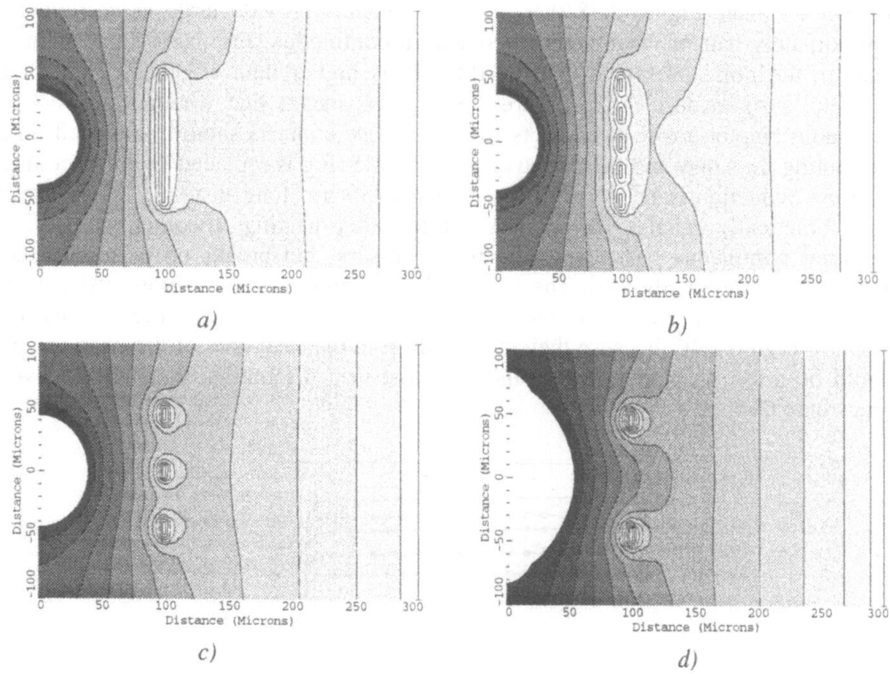

Figure 2.38

Constant voltage contour lines at the surface of the fragment simulated, when replacing the guard ring (a) by 5 (b), 3 (c) or 2 (d) contacts. The values of the voltages in the contour lines are the same in all cases. Observe that with 5 contacts (b) voltage (noise) levels at the right of the contacts are very similar to those obtained with a continuous ring (a).

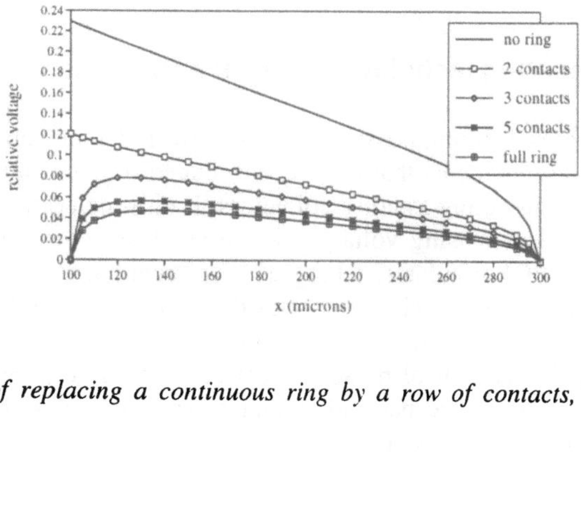

Figure 2.39

The effect of replacing a continuous ring by a row of contacts, in a lightly doped substrate.

We can see from Figure 2.38 that just two contacts provide large noise reduction, approximately half of what is achieved with a continuous ring. Nevertheless, in this situation the noise present at axis y=100 μm is higher than at other points of the substrate, as is made clear in Figure 2.38.d. This means that whether we choose a continuous ring or a row of contacts, some of these contacts should be placed on the line joining the noisy and the sensitive point. If the ring is replaced by three contacts, the noise reduction is better, and five contacts (50 μm long in total), provide noise levels practically equal to those with a full (100 μm) ring. In other words, in P- substrates continuous guard rings are not necessary, but breaks up to some tens of microns long can be allowed. The greater the discontinuities, the more efficacy lost. On the contrary, in P+ wafers the importance of these discontinuities is larger, as shown in Figure 2.40, because they mean a drop in the total area of the contacts. This should be not a problem, because the area lost in a discontinuity can be recovered somewhere else.

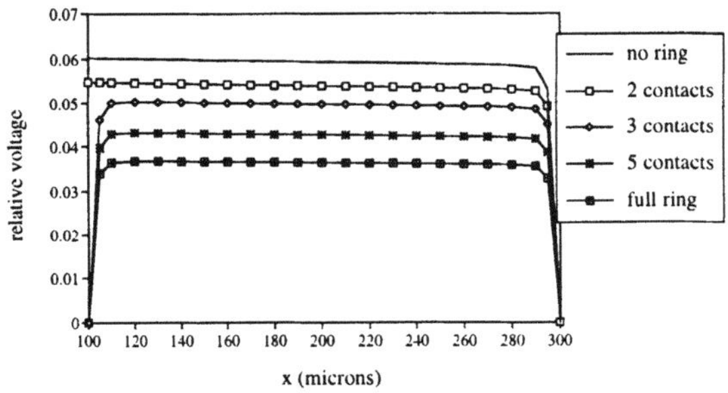

Figure 2.40

The effect of replacing a continuous ring by a row of contacts, in a heavily doped substrate.

2.5 Noise attenuating techniques in ideal conditions

This section compares different possible measures for reducing substrate coupling. All of these measures are elements that can be introduced to avoid noise transmission to sensitive parts, or items acting as noise collectors. It must be pointed out that the study is made assuming ideal biasing voltages, i.e. free of switching noise. The study is made with the MEDICI simulator, taking as a reference a structure with two MOS devices 10 μm apart, with the characteristics defined in section 2.2. As in that case, one of the transistors belongs to a switching inverter, while the other one is used as a sensor, measuring the amount of noise present at its drain terminal. The substrate is always biased with a grounded backside contact, wafer thickness being 500 μm. The bar diagram in Figure 2.41 shows the results obtained.

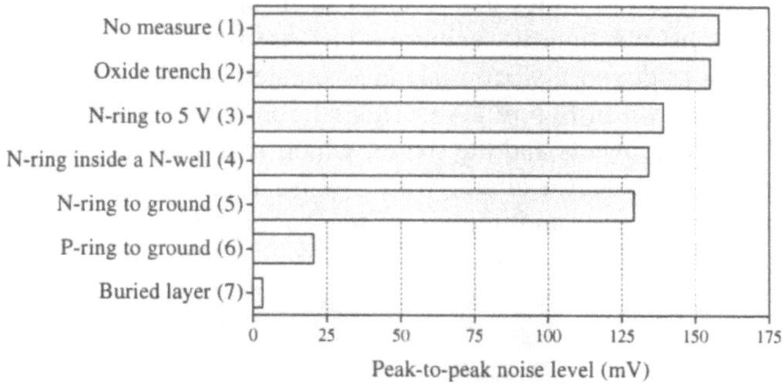

Figure 2.41

Comparison between the techniques studied for reducing substrate coupling.

The effect of placing protections to increase the insulation between coupled transistors is first studied. We start by considering the effect of placing an oxide trench across the noise path. This should be an effective solution whenever noise flows mainly along a depth smaller than that of the trench. If a trench 2 μm deep and 2 μm wide is placed between both NMOS, insignificant noise reduction is obtained, as can be seen by comparing bars (1) and (2) in Figure 2.41. This means that the depth of the trench should be much larger to become effective, which would make it difficult to implement and even less attractive. The effectiveness of oxide trenches is also evaluated in [20], reaching the same conclusion. An alternative would be also to include an oxide layer under the devices, that is to say, a silicon-on-insulator technology (SOI), so the noisy and sensitive devices are placed in fully insulated "silicon pockets", as shown in Figure 2.42. This technology is studied in [10], [20] and [21]. In these last two references an analysis is made according to frequency, reaching the same conclusion: the insulation achieved with SOI technologies is excellent at low frequencies, but they lose their effectiveness as the frequency is increased, due to capacitive effects. Substrate noise has frequency components in the range of hundreds of megahertz, and thus SOI technology is little attractive both due to its cost and to its limited capacity for reducing noise.

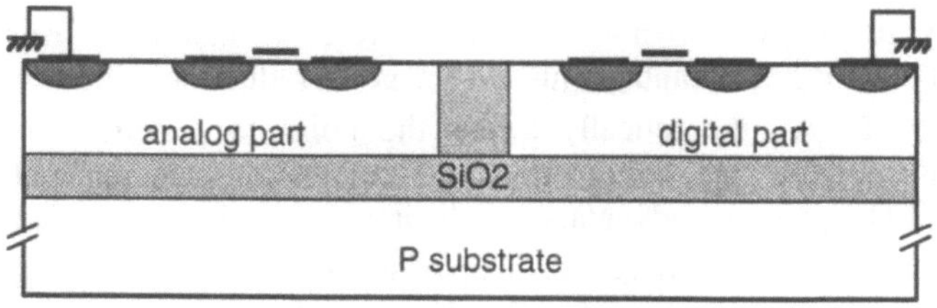

Figure 2.42

Silicon-on-insulator technology (SOI) for eliminating substrate coupling.

In connection with technologies that attempt to insulate the noisy and sensitive parts completely, we must mention the technique proposed in [22], consisting of an etching from the chip backside, so a silicon island is created in which the analog circuitry is implemented, as shown in Figure 2.43. This silicon island is joined to the rest of the IC through the interconnects and the oxide, which at the same time acts as an etching stopper. The authors report a circuit with a noise level in the analog section 30 dB above the residual noise, and they manage to eliminate it completely with this technique.

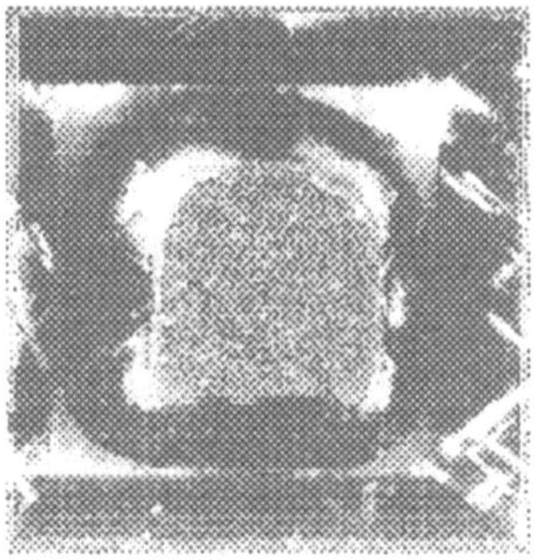

Figure 2.43

Image of the backside of an IC in which the technique proposed in [22] for reducing substrate coupling has been implemented.

A much more common technique for insulating the digital and the analog section involves placing a biased implant in between, i.e. a guard ring, that may fully surround the analog part. If the implant is a P type, the ring should be connected to ground, thus acting as a substrate bias contact. If the implant is an N type, it may be connected to *GND* or to V_{dd}. There is also a possibility of including an N+ ring inside an N-well. Usually two rings are placed, one P+ type and the other N+, to be sure of the efficacy. This is for example done in [23], with all implants grounded. On the other hand, in [24] an N+ ring connected to 5 V is used, inside an N-well, flanked by two P+ rings. There is no clear idea of which is the best option, even if some of the rings could be left out. Thus, we performsimulations of each of these measures separately, to verify their efficacy. It is firstly shown –bar (3) in Figure 2.41– that an N+ ring connected to V_{dd} does not offer much advantage, just a 12% noise reduction in our case. Placing the ring inside an N-well theoretically forces the noise to penetrate deeper into the substrate, but in Figure 2.41, bar (4), it can be seen that the difference obtained is very slight indeed. The potential advantage of placing an N+ ring connected to V_{dd} is that of obstructing the low impedance path provided by the channel stop layer usually implemented under the field oxide. The N+ ring forms a reverse-biased junction creating a depletion region impeding the flow of carriers. Anyway, the effect of this layer is of little relevance due to its shallowness and moderate conductivity. Two simulations of coupled devices were performed, one with the presence of a channel-stop layer with a doping level an order of magnitude higher than the bulk, and the other one without the layer. The noise level difference is just 4%, 155 against 149 mV_{pp}, respectively. The low efficacy of the N+ ring can thus be understood. Even if it is grounded, as in bar (5), better results are obtained, in spite of reducing the depletion layer.

A P-type ring not only improves substrate biasing, but also acts as a disturbance collector to ground. If the contact is connected to an ideal ground line, as in our simulations, the effects are especially beneficial. Though no ideal supply lines will exist in the real situation, and these may even inject noise, it is nevertheless good to know how much noise reduction could be obtained if conditions close to ideal were achieved. In our simulation, noise reduction obtained with a P-ring is spectacular, because in the reference case the substrate was only biased with a backplane 500 μm deep, so the effect of a surface contact near the transistors is considerable. In any event, it is clear that this type of ring is the only sort that can provide significant noise reduction. In [20] and [25] rings N+ and P+ are also compared, again reaching the conclusion that an N+ ring does not offer any advantage over a P+ one, which is the sort that really reduces substrate coupling.

Lastly, the results obtained when using a grounded heavily doped buried layer are shown. More specifically, a layer between 4 and 8 μm depth is assumed, with a 10^{19} cm^{-3} peak doping, extending all along the device simulated. The noise level thus obtained is the lowest achieved so far, just 2% of the original one. Of course, the effectiveness of the layer will depend both on its technological characteristics –square resistance–, and on the proximity and depth of the biasing contacts. This means that to guarantee its efficacy a technology allowing the implementation of deep contacts would be required. We should also remember that backside contacts in P+ wafers do not reduce high frequency noise due to skin-effect, which will not occur in a shallow buried layer. In chapter 5 the characteristics required to get effective buried layers are studied in detail.

2.6 Conclusions

From the results obtained in this chapter, a first set of conclusions can be drawn concerning the best techniques for eliminating substrate noise. Nevertheless, it must be stressed that the conclusions obtained are not definitive, but subject to the effect of packaging and noise at power-supply lines, which will be analyzed in the following chapters. This analysis in ideal conditions was necessary to evaluate the potentiality of each technique. In the event that effectiveness of the techniques should be lost when considering other parasitic effects, now it is at least known what can be achieved if measures for eliminating such parasitic effects are implemented.

Firstly, the characteristics of the wafers (P- or P+) and the presence of backside contacts completely determine the way noise is propagated and hence the best strategy for reducing it, so in practice this involves coping with two different problems.

P- wafers have uniform doping, which makes noise attenuate linearly with distance. This attenuation is due to the return path provided by substrate contacts, so attenuation is stronger the lower the impedance of this return path. The presence of a backside bias contact increases noise attenuation with distance, although given the usual wafer thickness and resistivity values its effect is not very noticeable in practice. For the same reasons, the resistance between the backside contact and the circuit surface is large enough to make it unsuitable for biasing the substrate, consequently losing all its usefulness. It must be anyway clear that simply separating two coupled circuits may be enough to eliminate substrate coupling, but the effectiveness of the separation depends on the impedance to ground of the contacts encountered along the propagation.

A guard ring is simply a low impedance contact in the disturbance path, providing a return path to ground and thus preventing noise from reaching the sensitive circuitry. The ring must be of the same type as the substrate (P+ ring in P wafers, N+ ring in N wafers). Implants of other types have proved to be inefficient because the noise that flows along the upper part of the wafer, including the channel-stop layer, is only a fraction of the total noise. It is not necessary to have a ring fully surrounding the sensitive circuitry. The disturbance propagates mainly along the straight line between coupled devices, so a ring in the form of a barrier across this path is enough. Making it longer is useless, because the effect of the ring decreases the further it goes from that line. As for the relative position of the ring, in practice it is not worth considering this, and better to account for the width of the ring. An interesting effect is that small openings in the ring –up to a few tens of microns– do not have a significant influence on its effectiveness. These openings can be important to solve connectivity or routing problems, and it is better if such openings are not in the straight line between coupled devices.

P+ substrates are not uniform, but present a lightly doped epi layer at the top of the wafer, a few microns thick. Noise originated by a disturbing element crosses the epi layer vertically, and is propagated to the whole chip through the high conductive substrate. Noise is not attenuated, and in these substrates the distance between devices or contacts is not relevant. There is attenuation if the bulk backside is connected to ground, but in this case noise is theoretically eliminated, because of the low impedance to ground of the highly conductive bulk. Nevertheless, the technological characteristics of the wafer –thickness, conductivity–,as well as the inductance of the package pins connected to that contact determine the value of this impedance and the real noise reduction. Moreover, for the usual wafer conductivities and thicknesses, the skin effect appears at frequencies around a gigahertz, depending on the conductivity, and then a contact at the chip bottom does not affect noise frequency components above this limit.

Given that in this type of substrate noise flows along the highly conductive bulk, a few microns below the surface, the effect of guard rings is lower than in P- substrates. On the contrary, noise non-dependence on distance now makes ring dimensions important. This means that in this case we can say that the larger the ring –whether longer, wider, or fully surrounding the sensitive circuits–, the lower the noise, in an almost linear relation. This makes the effect of ring discontinuities important, but involves the advantage that they can be compensated for by increasing the total dimensions. To sum up, in P- substrates the effect of guard rings decreases as they become more distant from the line between coupled circuits, while this is not the case in P+ epi substrates.

References

[1] *The PN Junction Diode*, 2ª ed., Gerald W. Neudeck, Addison Wesley Publishing Co., 1989.

[2] *Simultaneous Switching Noise of CMOS Devices and Systems*, R. Senthinathan, J.L. Prince, Kluwer Academic Publishers, 1994.

[3] A. Pun, T. Yeung, J. Lau, F.R. Clement, D. Su, "Experimental Results and Simulation of Substrate Noise Coupling via Planar Spiral Inductor in RF IC's", *Proc. IEEE Int. Electron Devices Meeting*, 1997.

[4] T. Yeung, A. Pun, Z. Chen, J. Lau, F.J.R. Clément, "Noise Coupling in Heavily and Lightly Doped Substrate from Planar Spiral Inductor", Proc. IEEE International Symposium on Circuits and Systems, pp. 1405-1408, June 1997.

[5] N.K. Verghese, D.J. Allstot, M.A. Wolfe, "Verification Techniques for Substrate Coupling and their Application to Mixed-Signal IC Design", *IEEE J. Solid-State Circuits*, vol. 31, no. 3, pp. 354-365, March 1996.

[6] *Principles of CMOS VLSI Design: A Systems Perspective*, 2nd edition, Neil H.E. Weste, Kamran Eshraghian, Addison-Wesley, 1992.

[7] M. Fukuma, W. W. Lui, "MOSFET Substrate Current Model Including Energy Transport", *IEEE Electron Device Letters*, vol. 8, no. 5, pp. 214-216, May 1987.

[8] T. Kawahara et al., "Substrate Current Reduction Techniques for BiCMOS DRAM", *IEEE J. Solid-State Circuits*, vol. 24, no. 5, pp. 1381-1388, October 1989.

[9] J.E. Chung et al., "Low-Voltage Hot-Electron Currents and Degradation in Deep-Submicrometer MOSFET's", *IEEE Trans. Electron Devices*, vol. 37, no. 7, pp. 1651-1656, July 1990.

[10] R.B. Merrill, W.M. Young, K. Brehmer, "Effect of Substrate Material in Mixed Analog/Digital Integrated Circuits", *Proc. IEEE International Electron Devices Meeting*, pp. 433-436, San Francisco, December 1994.

[11] *MEDICI, a two dimensional device simulator program*, Technology Modeling Associated, Inc. (TMA), Palo Alto, CA, 1993.

[12] M.J. Loinaz, D.K.Su, B.A. Wooley, "Experimental Results and Modeling Techniques for Substrate Noise in Mixed-Signal Integrated Circuits", *Symposium on VLSI Circuits Digest of Technical Papers*, pp. 40-41, 1992.

[13] D.K. Su, M.J. Loinaz, S. Masui, B.A. Wooley, "Experimental Results and Modeling Techniques for Substrate Noise in Mixed-Signal Integrated Circuits", *IEEE J. Solid-State Circuits*, vol. 28, no.4, pp. 420-430, April 1993.

[14] B.R. Stanisic, N.K. Verghese, R.A. Rutenbar, L.R. Carley, D.J. Allstot, "Addressing Substrate Coupling in Mixed-Mode IC's: Simulation and Power Distribution Synthesis", *IEEE J. Solid-State Circuits*, vol. 29, no. 3, pp. 226-237, March 1994.

[15] X. Aragonès, F. Moll, M. Roca, A. Rubio, "Analysis and Modelling of Parasitic Substrate Coupling in CMOS Circuits", *IEE Proceedings-Circuits Devices & Systems*, vol. 142, no. 5, pp. 307-312, October 1995.

[16] J.M. Casalta, X. Aragonès, A. Rubio, "Substrate Coupling Evaluation in BiCMOS Technology", *IEEE J. Solid-State Circuits*, vol. 32, no. 4, pp. 598-603, April 1997.

[17] H. Hasegawa, M. Furukawa, H, Yanai, "Properties of Microstrip Line on Si-SiO_2 System", *IEEE Trans. Microwave Theory Tech.*, vol. 19, no. 11, pp. 869-875, November 1971.

[18] *Fields and Waves in Communication Electronics*, S. Ramo, J.R. Whinnery, T. van Duzer, John Wiley & Sons, 1984.

[19] *RAPHAEL, an Interconnect Analysis Program*, Technology Modeling Associated, Inc. (TMA), Palo Alto, CA, 1994.

[20] K. Joardar, "A Simple Approach to Modeling Cross-Talk in Integrated Circuits", *IEEE J. Solid-State Circuits*, vol. 29, no. 10, pp. 1212-1219, October 1994.

[21] I. Rahim, B. Hwang, J. Foerstner, "Comparison of SOI versus Bulk Silicon Substrate Crosstalk Properties for Mixed-Mode IC's", *Proc. IEEE International SOI Conference*, pp. 170.171, 1992.

[22] P. Basedau, Q. Huang, "A Post Processing Method for Reducing Substrate Coupling in Mixed-Signal Integrated Circuits", *Proc. Symposium on VLSI Circuits*, pp. 41-42, 1995.

[23] M. Zuffada, R. Alini, M. Demicheli, M. Gregori, D. Moloney, S. Portulari, F. Sacchi, S.O. Arf, V. Condito, R. Castello, "A Single-Chip 9-32 Mb/s Read/Write Channel for Disk-Drive Applications", *IEEE J. Solid-State Circuits*, vol. 30, no. 6, pp. 650-659, June 1995.

[24] J. Eklund, R. Arvidson, "A Multiple Sampling, Single A/D Conversion Technique for I/Q Demodulation in CMOS", *IEEE J. Solid-State Circuits*, vol. 31, no. 12, pp. 1987-1994, December 1996.

[25] N.K. Verghese, D.J. Allstot, S. Masui, "Rapid Simulation of Substrate Coupling Effects in Mixed-Mode ICs", *Proc. IEEE Custom Integrated Circuits Conference*, pp. 18.3.1-18.3.4, 1993.

[26] T. Gabara, "Reduced Ground Bounce and Improved Latch-up Suppression through Substrate Conduction", *IEEE J. Solid-State Circuits*, vol. 23, no. 5, pp. 1224-1232, October 1988.

3 Substrate coupling modeling

This chapter reviews the different methods put forward in the literature for modeling the substrate. The first methods started from a very detailed description of the substrate and the layout, which allowed accurate and flexible modeling of all kinds of technologies and devices, and simple equivalent models were then obtained by numerical techniques. These methods are computationally very demanding, and some of them even follow the pattern of conventional device simulators. Successive model generations have added certain simplifications and have included pre-processing techniques to allow reducing computation requirements, but in spite of this fact such models do not turn out to be very efficient in very complex circuits. In the last few years simpler and less demanding modeling techniques have been sought for. Most of these methods are based on modeling the substrate as a mesh of resistances between neighboring devices or contacts. These resistances are obtained in many cases from semi-empirical expressions that make an estimation from geometrical parameters and technology-dependent fitting constants.

3.1 Substrate modeling by numerical methods

3.1.1. Methods based on solution of Poisson's equation

3.1.1.a. Modeling substrate behavior.

The first work in which a substrate model was obtained from a numerical resolution of Poisson's equation was developed by the Carnegie-Mellon University and published in 1993 [1], although some years before IBM had already done some preliminary work [2]. In its more complete development, the method starts with Poisson's equation (3.1) and continuity equations for both kinds of carriers (3.2) and (3.3).

$$\nabla^2\phi = -\frac{\rho'}{\varepsilon} \qquad \textit{Eq. (3.1)}$$

$$\frac{\partial n}{\partial t} - \frac{1}{q}\nabla J_n = -R \qquad Eq.\ (3.2)$$

$$\frac{\partial p}{\partial t} - \frac{1}{q}\nabla J_p = -R \qquad Eq.\ (3.3)$$

Expressing current in terms of electric field (neglecting diffusion current), differentiating (3.1) respect to time, and subtracting equations (3.2) and (3.3) yields

$$\varepsilon \frac{\partial}{\partial t}(\nabla E) + \frac{1}{\rho}\nabla E = 0 \qquad Eq.\ (3.4)$$

where E is the electric field applied, ρ is the substrate resistivity, and ε is its dielectric constant. This is the equation that models the substrate behavior. This expression can also be reached from Maxwell's equations, by leaving out the electric fields and using the identity $\nabla(\nabla \times a) = 0$. The goal will then be to solve equation (3.4) in the substrate under particular contour conditions. Given that an analytical solution is impossible if the presence of wells and layers with different geometry is considered, substrate must be discretized and the electric field at every point must be obtained by numerical methods.

3.1.1.b. Substrate discretization.
The numerical method most commonly used to solve substrate equations from its discretization is the finite difference method. A grid is generated and substrate is divided into a number of nodes, each them being the center of a cell like the one represented in Figure 3.1. The substrate must be uniform, with no concentration gradient, in this cell of volume Ω_j.

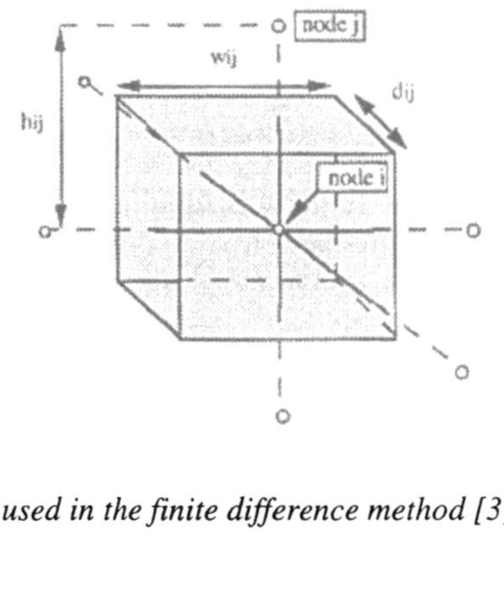

Figure 3.1

Integration volume used in the finite difference method [3].

From Gauss' law (3.5), by integrating over Ω_i, applying the divergence theorem, and assuming a constant field in the volume Ω_i, an expression of ∇E (3.6) in the aforementioned volume can be obtained:

$$\nabla E = \frac{\rho'}{\varepsilon} \qquad \text{Eq. (3.5)}$$

$$\nabla E = \frac{1}{\Omega_i} \sum_i E_{ij} w_{ij} d_{ij} \qquad \text{Eq. (3.6)}$$

By expressing E_{ij} as a function of the voltage difference between consecutive nodes, $E_{ij} = \frac{V_i - V_j}{h_{ij}}$, and substituting in (3.4), the equation determining the substrate behavior around node i in terms of lumped circuit elements is yielded,

$$\sum_j \left[\frac{(V_i - V_j)}{R_{ij}} + C_{ij} \left(\frac{\partial V_i}{\partial t} - \frac{\partial V_j}{\partial t} \right) \right] = 0 \qquad \text{Eq. (3.7)}$$

where $R_{ij} = \rho h_{ij} / w_{ij} d_{ij}$ and $C_{ij} = \varepsilon w_{ij} d_{ij} / h_{ij}$. This equation is equivalent to Kirchoff's current law at node i, which is linked to the neighboring nodes through an equivalent circuit like the one shown in Figure 3.2. This means that it has been shown that the substrate can be modeled as a mesh of resistances and capacitances. This equivalent circuit may be included in a mixed-signal circuit simulation, so currents through the substrate can be taken into account. The mesh nodes geometrically

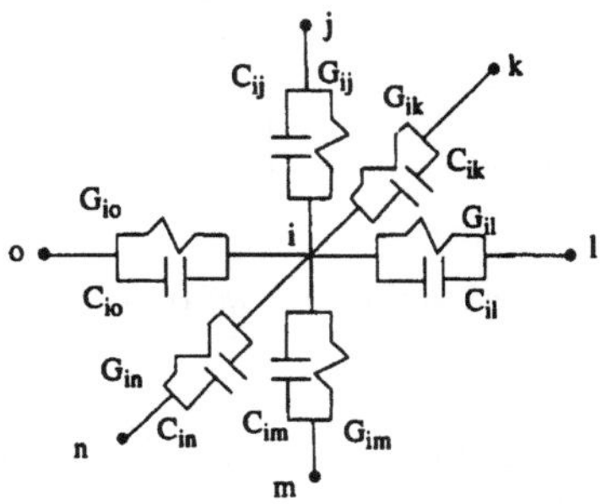

Figure 3.2

Substrate circuit model as a mesh of resistances and capacitances [3].

coincident with substrate contacts have their potential determined by the voltage applied to that node –the Dirichlet contour condition–, like the nodes coincident with device locations, which will be connected to their substrate terminals. Neumann contour conditions are applied to the other boundaries of the discretized region, consisting of forcing zero current entering or leaving the contour,

$$\frac{\partial \phi}{\partial n} = \frac{\partial J}{\partial n} = 0 \qquad Eq.\ (3.8)$$

where n stands for the contour normal direction.

Inserting the full substrate model mesh in a circuit simulator is not feasible due to the huge number of nodes that would be needed to discretize a realistic integrated circuit. In fact, the first difficulty when discretizing is to determine the mesh density in each region to achieve a minimum number of nodes. Some discussions [3] have sought to determine this mesh density. A technique proposed in [2] consisted in dividing the substrate into a number of nested cells with different densities, as represented in Figure 3.3. An equivalent admittance matrix as a function of the surface nodes is found for every cell, so the process may be iterated until only the nodes on the surface of the discretized volume remain.

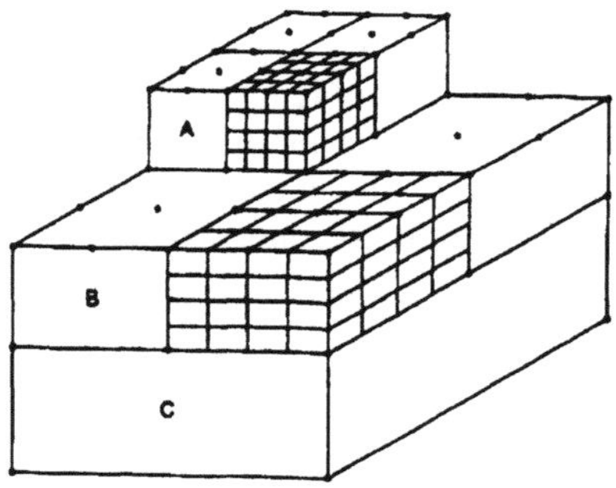

Figure 3.3

Cell nesting with different mesh densities, according to the technique proposed in [2].

In any event, these reduction techniques are still insufficient when dealing with very complex circuits. It seems clear that the maximum optimization will be reached when the substrate macromodel includes only the nodes connected to layout elements, whether these be devices or contacts. In other words, a mesh of equivalent admittances between every item in a layout is needed. These items will be called *ports* from now on.

3.1.1.c. Obtaining a substrate macromodel.

In the relevant literature two main techniques are presented for obtaining substrate macromodels, one based on AWE approximation (Asymptotic Waveform Evaluation), and the other based on a resistive approximation of the mesh of admittances, valid for frequencies up to a few gigahertz.

The AWE technique [4] allows approximating the behavior of linear circuits by using a few dominant poles and zeros in the time or frequency domain. Briefly, the technique is as follows: firstly, the moments of the function $V(s)$ are found, where $V(s)$ is the mesh voltage vector in the transformed domain, in response to a step excitation. From the initial condition ($t=0$) and steady-state vectors ($t\to\infty$), and by using a recursive algorithm, the moments may be found. From the full mesh moments, a reduction is made to obtain the macromodel moments, that consist only of the admittances of the n ports in the layout. With these moments, and by using Padé approximation, the poles and residues for every of the n^2 macromodel admittances are found. Usually, between two and four poles are enough to approximate every admittance. A detailed explanation of the algorithm will be found in [3].

As can be observed, the AWE technique is extremely complex and arduous. Although, once implemented, results can be obtained significantly faster than with a device simulator, it is nevertheless time-consuming because some large matrices must be solved during the moment calculation. An alternative to this technique is that of modeling the substrate by assuming static conditions. The validity of the model will be then restricted to the slow-wave behavior according to the classification made in chapter 2. In these conditions, equation (3.7) is reduced to the resistive terms, and can be directly obtained from Laplace's equation. This way provides a system of equations associating currents and voltages at every node as a function of a conductance matrix,

$$[G][V]=[I] \qquad \text{Eq. (3.9)}$$

Matrix [G] is sparse, with no more than seven non-zero elements in every row or column. System (3.9) must be solved for every port in order to obtain the admittance macromodel, forcing a unity voltage while maintaining zero-voltage at the rest of ports. The admittances between that port and the rest are equal to the inverse of the current in every port.

The problem with this last technique is the treatment of well junctions. The most common solution is to find an equivalent capacitance of the well-substrate junctions, and connect it to the macromodel through a single node [5]. This way wells are converted into another element of the circuit, and will be connected to a port of the substrate macromodel. The precision achieved with this treatment may become affected, especially if wells are large. One alternative could be to partition the well and connect it to the macromodel through different ports, but this increases the computational cost of both obtaining the macromodel and of the simulations. A similar technique could be used to include the parasitic capacitances of the interconnects to the substrate.

3.1.1.d. Matrix solution.

Both in the AWE technique and in modeling in static conditions, a key point determining the speed and feasibility of macromodel generation is the matrix equations solution. Direct techniques such as LU factorization are not applicable due to the large size of the matrices, and because this method does not take advantage of the regularity and symmetry of the matrices. On the contrary, iterative techniques such as the Gauss-Jacobi method, Gauss-Seidel, and Simultaneous Over-Relaxation preserve matrix sparsity throughout the solving process. Given that matrices are symmetric positive definite, the Incomplete Choleski Conjugate Gradient (ICCG) method is usually adopted. This method guarantees convergence in less than n iterations, where n is the order of the matrix. A more extensive review of these solution methods can be found in [3].

Substrate modeling from the solution of Poisson's equation with any of the options presented has been used as a basis for certain tools in which substrate effects were taken into account. In [5] a tool is presented –LAYIN™– which, by taking layout and technological data, obtains the substrate macromodel and allows its circuit simulation together with the rest of the IC. This tool has already been included in a commercial placement & routing software package [6]. In [7] and [8] a tool –RAIL– is presented that synthesizes the power-supply distribution net in an IC, and the pad placement, taking into account substrate noise minimization. The tool needs a substrate macromodel, which may be generated by the AWE technique. The same authors present another tool –WRIGHT– in [9] to perform the placement in a mixed-signal IC taking into account substrate noise coupling restrictions in the analog section. These techniques have even been used to analyze thermal substrate coupling [10]. Nevertheless, they are complex techniques and need large computation times. With AWE technique and using the ICCG method to solve the matrices, $2q+1$ matrix inversions are needed for every port to find the admittance values, where q is the number of poles chosen in the approximation. In the case of the resistance macromodel, the number of inversions is reduced to one per port. In any case, taking into account that a low-complexity circuit has thousands of gates, implementation of this technique is hardly practical.

3.1.2. Methods based on obtaining Green's function

An alternative method computationally less demanding than the finite difference methods is the Boundary Element Method (BEM). It is assumed that the substrate is composed of a number of stacked layers of uniform conductivity. This is an assumption which matches reality, except for a few microns under the silicon surface, where wells and channel stop layers may have irregular contours. For frequencies under a gigahertz, we know that the substrate behaves resistively, and that the potential distribution is given by Laplace's equation,

$$\nabla^2\phi = 0 \qquad \text{Eq. (3.10)}$$

It is assumed that a number of contacts are present on the contour of the volume to be analyzed (basically on the upper surface, except for a possible backplane), which will force a given potential in those areas. On the rest of the surface Neumann contour conditions are imposed (3.8). According to Green's theorem, an electrostatic potential at a point *r*, due to a unity current injected at point *r'*, is given by the following expression

$$\phi(r) = \int_{V'} J(r')G(r,r')dv' \qquad \text{Eq. (3.11)}$$

where *J(r')* is the injected current density, and *G(r,r')* is Green's function: the potential at a position *r* due to a point charge placed at *r'*. Given that current injection is performed on the ports, whether contacts or devices, and these may be considered two-dimensional, equation (3.11) can be reduced to a surface integral over the ports.

The first problem is to find the substrate Green's function given the dimensions of every particular case, with Neumann contour conditions. There are several methods for finding this function. The method of images is simple, but in non-uniform mediums it leads to an expression of *G(r,r')* in terms of nested infinite series. Furthermore, the finite substrate boundaries worsen the problem. The expansion of the domain eigenfunctions also yields a triple infinite series that converges slowly. The classic method of variable separation has successfully been used in treating a substrate with N layers, although it leads to complex expressions, in terms of a slowly converging double infinite series [11], [12]. The authors propose calculating it with the 2-D discrete cosine transform (DCT) of the series, which may be efficiently calculated with a fast Fourier transform (FFT). In [13] the equation obtained for the most typical case of two layers is shown, from separation of variables in cylindrical coordinates. The result obtained is a simple series but still of slow convergence, which must be truncated when enough precision is obtained. In order to reduce the number of contour conditions, their method assumes that substrate sidewalls are placed at infinity, and that substrate thickness is also infinite. The first assumption can be accepted insofar as devices are usually placed distant enough from the saw lane, but the second assumption is more questionable. Anyway, the effects of the chip sidewalls may be taken into account with the method of images, while the thickness can be accounted for if a backside contact is assumed. The same authors had previously obtained expressions for a two-layer substrate using the operator technique [14]. In this case Neumann contour conditions were applied at the sidewalls, but the result was again presented in the form of a double infinite series of difficult convergence.

After determining the Green's function with some of the former techniques, the next step is to solve (3.11), and obtain the impedance matrix. This is done numerically, by a contour discretization. Given that Green's function already took into account the substrate boundaries in its formulation, it will only be necessary to discretize the ports connecting the substrate to the rest of the circuit, which are divided into a number of panels, as shown in Figure 3.4. This way a system of equations relating potentials and currents in each of the discretized elements is yielded,

$$[\phi]=[Z][I] \qquad Eq.\ (3.12)$$

where each of the elements in the impedance matrix is, in its simplest formulation [15],

$$Z_{ij}=\frac{1}{S_i S_j}\int_{S_i}\int_{S_j} G(r,r')ds'ds \qquad Eq.\ (3.13)$$

where S_i and S_j are the surfaces of both panels. Once elements z_{ij} have been calculated, and the matrix Z has been evaluated, it is inverted to obtain the admittance matrix Y. Later this matrix Y may be treated to group the elements corresponding to the same port. An assignment of one panel per port leads to excessively large errors, because the current density is not uniform in the port. A discretization like the one shown in Figure 3.4.b is usually enough to get errors lower than 10%.

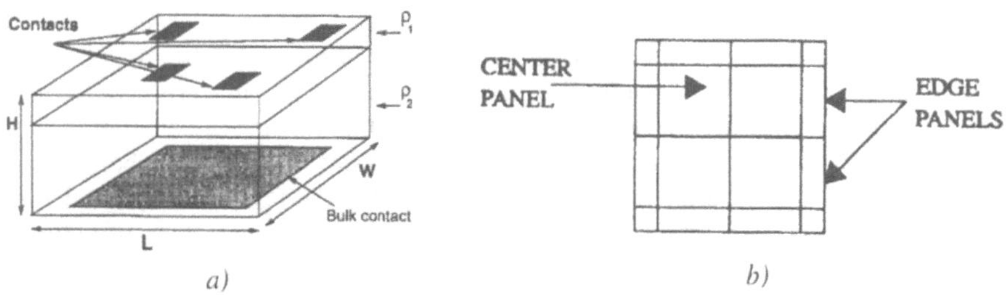

Figure 3.4

a) Substrate view with ports on the surface [13]. b) Port discretization into a number of panels. The area of the panels is different to take into account the non-uniform current density [15].

Apart from finding the Green's function with the desired contour conditions, the main problem with this method is the inversion of the impedance matrix. These matrices are much smaller than those appearing in the finite difference method presented in the former section, because in this case only the ports are discretized. Nevertheless, now the matrices are not sparse, so the methods exploiting the matrix symmetry will not be usable now. Using the LU method, the inversion is an $O(N^3)$ order problem, where N is the number of panels. This complexity is not acceptable for large circuits, with a huge number of ports. In [13] the Shur method is proposed to simplify the inversion, by making an approximation in which only the admittances between an element and other ports within a radius around it are computed. This way a sparse matrix is obtained, the process complexity now depending on the radius inside which admittances will be calculated. In a similar way, a partitioning algorithm is proposed in [11], where every port is surrounded by the closer ports, and further away by partitions of increasing size. An "equivalent port" is found for every partition, which will be used to take into account the influence of distant contacts in the computation of the admittance between two nearby ports. Only the admittances between ports within a single partition are computed, so again a sparse matrix is obtained. On the other hand, the partitioning algorithm must be applied N times. Lastly, in [15], [16] the GMRES method is used, which solves (3.12) iteratively, minimizing the error $v^k\text{-}Z{\cdot}i^k$, where k is the iteration index. The solution (3.12) is then $O(N^2)$, but can be reduced to O(N) with multipole expansions and local expansions for every panel.

In [15], [16] a method for avoiding computation of Green's function on the substrate is proposed. It consists in creating a data base of values of z_{ij} (3.13) according to the distance between panels. If the port discretization is done with a reduced set of panels, a set of z_{ij} for a range of distances and for a given technology may be obtained. When analyzing a particular situation, elements z_{ij} in the impedance matrix would be obtained from interpolation between the values stored in the data base. This way the evaluation of Green's function for every particular case would be avoided. The values of this data base could be obtained either by solving Green's function in a number of simple situations, or from measurements on test structures. The authors assume that the value z_{ij} is independent of the distance to the chip edges, which is demonstrated. The influence of close contacts on the evaluation of z_{ij} should also be verified, though.

The boundary element method has been used to implement some substrate macromodel extraction tools, such as SUBTRACT [16], or SPACE [17]. This last tool also uses the same method to extract capacitances between interconnects, from a 3-D analysis. Nevertheless, the complexity of the computations and the fact that once again they deal with matrices resulting from discretization, makes the extraction slow and unfeasible for circuits with a large number of ports. The main problem both in the BEM and in the finite difference method is that they analyze the substrate in the whole chip as a single unity, so in a realistic circuit they must deal with thousands of elements representing the thousands of contacts and devices. In the last few years, the trend seen in the same authors has led to obtaining macromodels in much more approximate ways, finding the impedance between pairs of elements from pseudo-empirical expressions, and abandoning the numerical methods that analyze the substrate as a whole [18]. These methods, with concepts utterly unlike the others, are presented in the following section.

3.2 Substrate modeling with approximate methods

3.2.1. Substrate noise evaluation from signature superposition

The first of the methods presented in this section is original and different to the rest in its concept [19]. The objective is to approximate noise injectors with one or more current sources, and use the superposition principle to compute total noise.

The first step is to characterize substrate noise injection. It is assumed that noise is introduced to the substrate through two mechanisms: impact ionization and capacitive coupling. This means that the noise introduced through the substrate contacts isignored. Considering these two noise sources, the noise signature of each of the cells in a digital library is determined. Each signature is obtained from a layout extraction, and by applying all possible input combinations. The noise characteristics of every cell are stored in a table of signatures. As for propagation, a resistive behavior is assumed for the substrate, so there will be no time skew in the noise received. Furthermore, it is assumed that noise attenuation is not very dependent on the distance, which in fact only occurs in heavily doped substrates, or if all noise sources are placed at a similar distance to the sensitive node.

When evaluating noise in a given circuit, simultaneously switching cells with identical signatures are grouped together. A classic gate delay model is included in the simulator, consisting of a gate intrinsic component plus a load dependent component, which includes the interconnect capacitances. Then the total noise in a substrate point is determined, first by convoluting every noise signature with the activity of the associated gates, and afterwards by applying the superposition principle to add the noise produced by every group of cells with identical signatures. The authors compare the results obtained, expressed as a noise spectrum estimation, against the results obtained with the BEM [11], and check the validity of their approximation.

As for superposition, an estimation of the noise attenuation in the substrate is made. The attenuation coefficients are actually dependent on the technology and the layout, and should be estimated by some of the numerical methods presented in the former section. In any case, although the starting concept is attractive, a number of approximations are made that limit the validity of the method, although most of them can be overcome. The only limitation that is hard to get over is that of ignoring the noise introduced by the contacts, which must be included in a realistic simulation, as will be seen in the next chapter.

3.2.2. Model obtaining by Delaunay triangulation

The method presented next [20] seeks to obtain a substrate macromodel avoiding an analysis that considers the whole substrate. The model contains approximations for the impedance between close ports, ignoring the impedance between distant ports or between ports with other elements in the middle. That is to say, the impedances can be individually obtained from geometrical data, ignoring the other elements. A procedure called Voronoi tessellation is used to discretize the layout. In this procedure every port is a node of the mesh discretizing the substrate. For grater accuracy, it is useful to add nodes in the layout regions free of ports, near the wells, and on the chip boundaries. Around every node an irregular polygon is created (i.e., the discretization is not rectangular, as occurred in the methods seen so far), and this polygon encloses the area closer to the node than to the neighboring nodes. The resulting polygon mesh is called Voronoi diagram, and the dual of this mesh, i.e. the set of lines between nodes in adjacent polygons, is called Delaunay triangulation. The macromodel that will be generated only includes the impedances between the nodes connected after a Delaunay triangulation. Figure 3.5 shows the Voronoi diagram for a number of point ports, and the corresponding Delaunay triangulation on its right. The Voronoi diagram generation can be performed with a algorithm of complexity O(NlogN), where N is the number of nodes in the mesh.

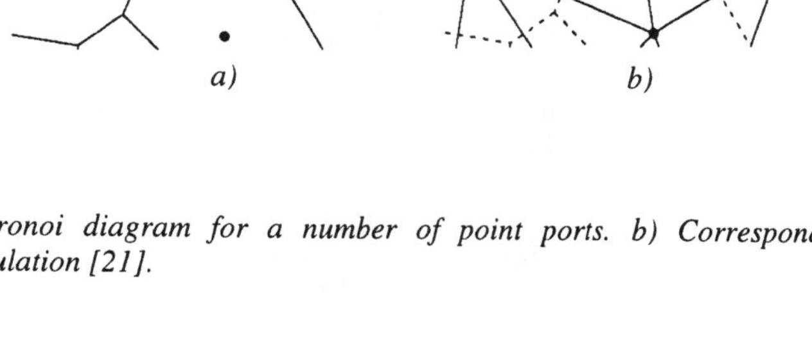

Figure 3.5

a) Voronoi diagram for a number of point ports. b) Corresponding Delaunay triangulation [21].

Once the layout has been discretized, the substrate is divided in its vertical direction into a number of layers of thickness depending on the depth and carrier concentration gradient. This way every Voronoi polygon is in fact turned into a Voronoi volume, in which a constant electric field is assumed. The resistance between the corresponding node and one of the edge faces is approximated by

$$R = \frac{1}{\sigma A} \qquad \textit{Eq. (3.14)}$$

where l is the distance between the node and the face, σ is the average conductivity in the volume, and A is the area of the face. By adding the resistances between adjacent volumes, which are connected in series, a three-dimensional mesh of resistances is obtained as a substrate macromodel. The nodes separated by a well junction are connected through a capacitance, modeled with expression (2.1). The authors validate the method by comparing the results against those obtained from MEDICI simulation of some simple situations.

Although the mesh obtained is simpler than the one generated with a rectangular discretization, because now there are only nodes at the points of interest, the circuit obtained is still too complex, because there are still more nodes than ports. The same authors, in [21], propose a method for reducing the mesh with a technique similar to AWE, mentioned in the former section. The technique proposed requires the inversion of sparse matrices, ultimately a process of similar complexity to the finite difference method presented in 3.1.1 is obtained, simply with smaller matrices in this case.

In [18] a modeling technique of P+ substrates is presented, also based on Delaunay triangulation. In this case, given that the highly conductive bulk behaves electrically as a single node, the mesh of resistances is reduced to a surface mesh, with resistances between every port, plus the resistance between every port and the heavily doped part, i.e. the resistance across the P- epi layer. The values of the resistances are found by interpolating known values for a given technology. These known values may be obtained either experimentally or by one of the methods presented in section 3.1. From the interpolation curves, the following expressions for the model resistances may be obtained,

$$R_{vert} = \frac{1}{k_1 + k_2 P + k_3 A} \qquad \textit{Eq. (3.15)}$$

for the vertical resistance through the epi layer, where P is the perimeter of the port, A is its area, and k_1, k_2 and k_3 are fitting constants, and

$$R_{horz} = \frac{K d^{p}}{\sqrt{A_a} + \sqrt{A_b}} \qquad \textit{Eq. (3.16)}$$

for the resistance between ports, where d is the distance between them, A_a and A_b are the port areas, and p and K are again fitting constants. The former expressions must be modified empirically when ports are very close, because the current fraction crossing the epi layer is then smaller.

In all the proposals seen so far the impedance between two ports is modeled as a single resistance. In [22], [23] a more complex model is put forward, also valid for frequencies over the limit of the slow-wave behavior, up to 40 GHz. The model is composed by a mesh of resistances and capacitances in parallel. The authors propose to determine the elements of this equivalent circuit by means of numerical simulation.

Lastly, in [24] an added simplification to model P+ substrates is put forward. Given that in mixed-signal circuits the digital part is the largest one, they suggest including in the model only the resistances associated with the transistors of the highest frequency part, which is the one that generates the highest noise. Transistor speed or size is in fact more important than switching frequency in noise generation.

3.2.3. Expressions for calculating the resistance between two ports

Just as in [18] the authors propose semi-empirical expressions to determine the resistances in their model, other authors have proposed expressions to approximate the resistances of the substrate macromodels. The first approximation is found in [25], where the resistance between two circular contacts is determined by means of an integration algorithm, assuming a semi-infinite space. The expression obtained is

$$R_{horz} = \frac{\rho}{2r}\left(1 - \frac{2}{\pi} arcsin\left(\frac{r+D}{r}\right)\right) \qquad Eq.\ (3.17)$$

with ρ being the substrate resistivity, r the radius of the contacts, and D the distance between these. In [26] a calculation of the resistance between rectangular contacts is made, first by solving in two dimensions, and afterwards extending to three dimensions assuming that the resistivity does not vary in dimensions x and y (surface). The two-dimensional expression is obtained by solving Laplace's equation in a semi-infinite space, assuming a zero potential at an infinite depth, and applying Neumann conditions on the surface, except for the contacts. The expressions obtained are really complex, and the authors fail to show them in closed form equation, preferring to present graphic representations of the resistance dependence on certain geometrical parameters.

Later, and aiming substrate coupling simulation, an empirical expression for the resistance between a port and a heavily doped substrate is determined in [27]. Given that in P+ substrates the disturbance is mainly propagated through the P+ bulk, the authors ignore the direct resistance between ports, and model the substrate only with the vertical resistances, considering the P+ bulk to be a single node. The expression of this resistance is obtained as the parallel of the contribution of the perimeter of the port and its area,

$$R_{vert} = \left(\frac{k_1 \rho e}{(L+\delta)(W+\delta)}\right) \Bigg\| \left(\frac{k_2 \rho}{2(W+L+2\delta)}\right) \qquad Eq.\ (3.18)$$

where e is the epi layer thickness, L and W are the length and width of the port, and k_1, k_2 and δ are fitting parameters.

Lastly, in [28] an equivalent model is obtained for two coupled contacts, each of them surrounded by a guard ring, as depicted in Figure 3.6.a. The coupling model consists of a net of resistances like the one shown in Figure 3.6.b. Expressions are proposed for each of these resistances, which are generated from reasoning about how should they depend on geometrical dimensions. The values obtained from the expressions are not compared with simulations or measurements. The following expression is proposed for resistances R_1 and R_5,

$$\frac{1}{R} = a\left(\frac{L}{S}\right) + b\left(1 - \frac{c}{S}\right) \qquad Eq.\ (3.19)$$

while for resistances R_2 and R_4, the expression is

$$R = \left(\frac{d}{L} + e\right)(f - g\log W) + h\frac{S}{W} \qquad Eq.\ (3.20)$$

where L is the contact length, S its distance from the guard ring, W the width of the ring, and $a..h$ are fitting constants. For R_3 the approximation (3.17) for circular contacts is taken, adding another three fitting constants. The expressions proposed lack an analytic or experimental verification, and they need such a number of fitting

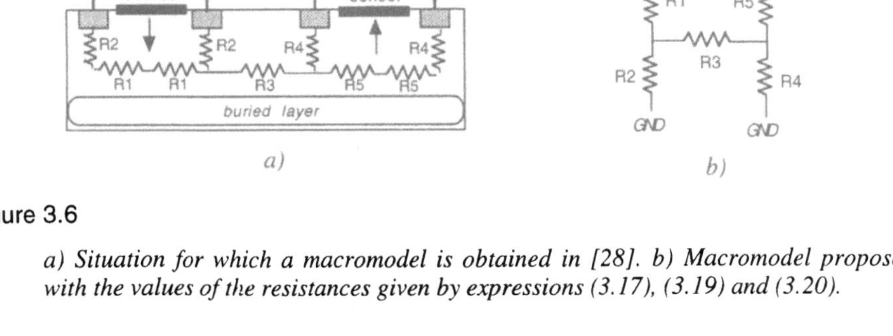

Figure 3.6

a) Situation for which a macromodel is obtained in [28]. b) Macromodel proposed, with the values of the resistances given by expressions (3.17), (3.19) and (3.20).

constants that they are little use. In the same paper similar expressions may be found to model coupling in a SOI technology.

3.2.4. Discussions on macromodel validity

Throughout this chapter the evolution of techniques proposed in the literature for modeling the substrate has been presented. Generally speaking, the first method pursued a substrate modeling from an exhaustive discretization of the bulk, using finite difference methods. This is the technique used by device simulators like MEDICI, and given its resolution it is only applicable to simple situations. The first goal was to obtain a substrate macromodel consisting of a mesh of resistances –for low frequencies, and discarding the treatment of wells– between all ports in a layout, so if it contained N ports, the macromodel would consist of $N(N-1)/2$ resistances. This macromodel could then be included in a circuit netlist, so a SPICE simulation could take into account coupling through the substrate. The problem is that, when analyzing real circuits with a very large number of ports, the number of resistances in the model makes it intractable. The objective has then become to obtain a model with a number of resistances linearly dependent on the number of ports N. Given that the resistance between distant ports is very high, macromodels considering only the resistance between adjacent ports have been proposed, so the N complexity goal is achieved. Last, in P+ substrates the mesh may be simplified because the highly conductive bulk behaves like a single node, and no discretization in the vertical direction is needed. At circuit level, a similar problem arises, because it is hardly feasible to perform a transient simulation of a complex digital circuit to account for the noise contribution of every single transistor. Instead, techniques like the signature superposition presented above [19], or the technique discussed in [29] for transient simulation of high-frequency circuitry have been proposed to evaluate the disturbance introduced by noisy sections.

This evolution towards more treatable substrate models has developed at the cost of generalization. So, while an exhaustive discretization allowed the simulation of any situation, considering the bulk as a single node is in theory only valid for P+ substrates, while it is not clear whether Delaunay triangulation can be applied to all situations.

We will next discuss what the limitations on the usage of two proposed macromodels are, and the error introduced by these macromodels in some simple situations. The simplest model considered is the one consisting of a mesh of resistances between neighboring elements, determined for example by Delaunay triangulation, plus a vertical resistance between every port and the bulk inside, considered as a single node. This model will be called *Π model*, because this is the shape adopted by the mesh of resistances in the elementary case of only two ports, and is shown in Figure 3.7.b. The second macromodel considered is the one consisting of the resistances extracted between all ports, adjacent or not, *–N(N-1)/2* resistances–. This macromodel will be called the *full model*, because it does not leave out any of the elements of which consists, and is shown in Figure 3.7.a. Remember that both the macromodels obtained from AWE and BEM methods are full models.

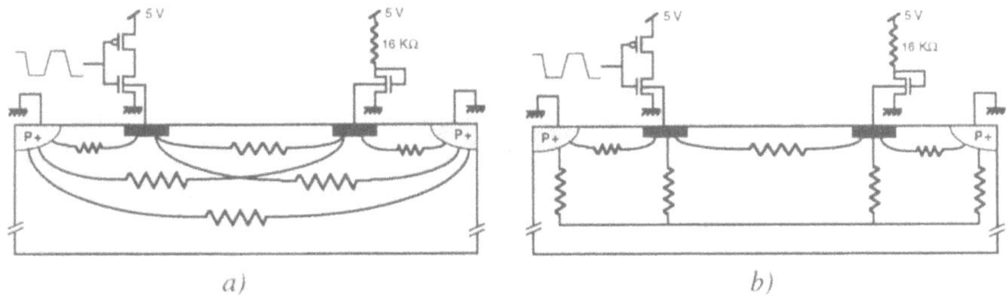

Figure 3.7

Substrate models considered. a) Full model –N(N-1)/2 *resistances–. b) Π model* –2N-1 *resistances, in a two-dimensional situation–.*

The main problem that may arise when using the *Π model* is when simulating the coupling between two devices with another element in between –a substrate contact, for example. In this case, the direct coupling resistance is not included in the model, and noise may be underestimated. In P+ substrates, though, given that noise propagates basically through the heavily-doped part, this is not usually a significant problem. This could be a problem when substrate is biased by means of a backside contact. If this contact was ideally grounded, the highly conductive bulk would also be well grounded, and simulation would show no noise transmission between coupled devices. If the backside ground contact is not ideal, the error will be smaller the higher the impedance to ground. Then, the error will be smaller in the worst situations. Also, when substrate is biased by means of a backside contact, currents are distributed along the bulk, which will be hardly modeled with lumped resistances. On the other hand, when there is no backplane, noise is propagated just under the epi layer, directly from the noisy transistor to other ports, and thus a model with lumped elements is more suitable. This may be checked in Table 3.I, which compares the simulation of a simple situation consisting of two coupled devices 30 μm apart–results for other distances showed similar errors. Simulations are made with a device simulator, with SPICE using a *full model* of the substrate, and with SPICE using a *Π model* of the substrate. In these last two cases, resistances were extracted from device simulation.

In P- lightly doped substrates, the *Π model* is clearly inadequate when there is a guard ring or any other item between the coupled devices. What happens in this case is something similar to what occurred in P+ substrates with a grounded backplane: the Delaunay triangulation does not take into account resistances between non-

Table 3.I.

Difference in results between the substrate models, and the device simulator, considering a P+ epi wafer.

P+ substrate:	device simulator	full model	error	Π model	error
2 contacts	48 mV	47.9 mV	0.21%	53.6 mV	11.6%
2 contacts + ring	28 mV	26.2 mV	6.4%	23.7 mV	15.4%
2 contacts + backplane	4.25 mV	4.73 mV	11.3%	5.06 mV	19%
only backplane	4.5 mV	5.33 mV	18.5%	5.33 mV	18.5%

consecutive elements, and the mesh connecting the coupled devices through the substrate inside has an excessively large impedance. In this situation, a model like the one presented in [21] would be more suitable, as it treats the substrate as a superposition of layers, and a low impedance path between devices can be found despite the presence of other elements in between. An alternative we propose is to include the direct coupling resistance to a *Π model* in these cases. This involves a problem for the methods based on Delaunay triangulation, as they seek to approximate the resistance between consecutive nodes with semi-empirical expressions or interpolation curves. Keeping the same philosophy would mean obtaining the semi-empirical expressions or interpolation curves for the three resistances that connect three nodes, taking into account all possible geometrical layouts, which is hardly feasible. A second problem that arises is that including a coupling resistance between non-consecutive nodes implies a model modification only where necessary, which is difficult to systematize, because previous information of the nature of every port – digital gate, sensitive transistor, contact– would be needed. An alternative method that does not need this information involved using *the full model*, and ignoring the resistances with values above a certain limit. This method has also been tested, and the model size is extremely low, being approximately linear with the number of nodes.

On the other hand, the *Π model* tends to overestimate the noise as the resistance to the substrate contacts increases. That is to say, the worse the substrate biasing, and the larger the noise, the worse the model.

Table 3.II now shows the error made with the SPICE simulation of two coupled transistors using the proposed macromodels, against the device simulation results. Results are now shown for two different distances *d* between devices –30 and 50 μm–, and different distances *c* between the devices and substrate contacts. It may be seen how unsuitable the *Π model* is for treating guard rings, but if the simulations are repeated, this time including the direct coupling resistance, the noise levels measured are 22 and 15.9 mV, instead of the 2.38 and 2.25 mV that appear in Table 3.II. This means that the errors are now 10% and 4.1%, results which are comparable to the other situations. The coincidence of waveforms obtained with MEDICI and the *Π model* can be observed in Figure 3.8, in both types of substrates.

Table 3.II.

Difference in results between the substrate models, and the device simulator, considering a uniform P- wafer.

P- substrate:	device simulator	full model	error	Π model	error
2 contacts					
d=30 μm, *c*=11.6 μm	56 mV	58.6 mV	4.6%	70.1 mV	25.2%
d=50 μm, *c*=3.6 μm	43 mV	39.5 mV	8.1%	44.55 mV	3.6%
2 contacts + ring					
d=30 μm, *c*=11.6 μm	20 mV	20.25 mV	1.25%	2.38 mV	88%
d=50 μm, *c*=3.6 μm	16.6 mV	15.1 mV	9%	2.25 mV	86.5%
2 contacts + *backplane*					
d=30 μm, *c*=11.6 μm	51 mV	51.7 mV	1.4%	62.1 mV	21.7%
d=50 μm, *c*=3.6 μm	38 mV	33.9 mV	10.8%	38.5 mV	13.2%

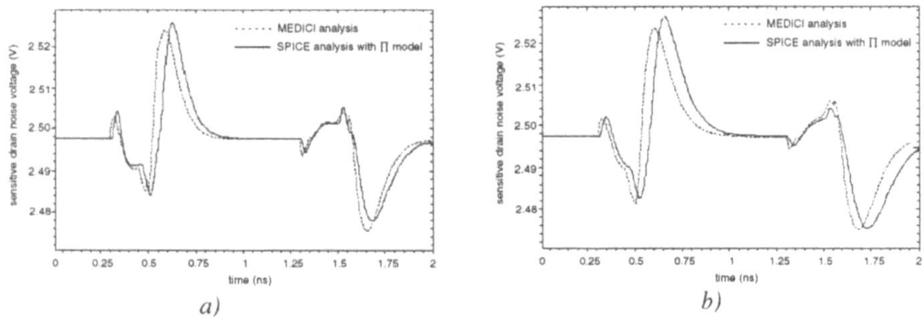

Figure 3.8

Noise waveforms obtained from MEDICI simulation and with the Π model. a) In a P+ substrate. b) In a P- substrate.

References

[1] N.K. Verghese, D.J. Allstot, S. Masui, "Rapid Simulation of Substrate Coupling Effects in Mixed-Mode ICs", *Proc. IEEE Custom Integrated Circuits Conference*, pp. 18.3.1-18.3.4, 1993.

[2] T.A. Johnson, R.W. Knepper, V. Marcello, W. Wang, "Chip Substrate Resistance Modeling Technique for Integrated Circuit Design", *IEEE Trans. Computer-Aided Design*, vol. 3, no. 2, pp. 126-134, April 1984.

[3] *Simulation Techniques and Solutions for Mixed-Signal Coupling in Integrated Circuits*, N.K. Verghese, T.J. Schmerbeck, D.J. Allstot, Kluwer Academic Publishers, 1995.

[4] L.T. Pillage, R.A. Rohrer, "Asymptotic Waveform Evaluation for Timing Analysis", *IEEE Trans. Computer-Aided Design*, vol. 9, no. 4, pp. 352-366, April 1990.

[5] F. Clement, E. Zysman, M. Kayal, M. Declercq, "LAYIN: LAYout INspection CAD tool dedicated to the parasitic effects through the substrate of integrated circuits", *Proc. Power and Timing Modelling and Optimization Workshop, PATMOS*, pp. 267-276, 1995.

[6] *Snake Technologies*, information available at http://www.snaketech.com.

[7] B.R. Stanisic, N.K. Verghese, R.A. Rutenbar, L.R. Carley, D.J. Allstot, "Addressing Substrate Coupling in Mixed-Mode IC's: Simulation and Power Distribution Synthesis", *IEEE J. Solid-State Circuits*, vol. 29, no. 3, pp. 226-237, March 1994.

[8] B.R. Stanisic, R.A. Rutenbar, L.R. Carley, "Addressing Noise Decoupling in Mixed-Signal IC's: Power Distribution Design and Cell Customization", *IEEE J. Solid-State Circuits*, vol. 30, no. 3, pp. 321-326, March 1995.

[9] S. Mitra, R.A. Rutenbar, L.R. Carley, D.J. Allstot, "Substrate-Aware Mixed-Signal Macrocell Placement in WRIGHT", *IEEE J. Solid-State Circuits*, vol. 30, no. 3, pp. 269-277, March 1995.

[10] N.K. Verghese, S. Lee, D.J. Allstot, "A Unified Approach to Simulating Electrical and Thermal Substrate Coupling Interactions in ICs", *Proc. IEEE International Conference on CAD*, pp. 422-426, 1993.

[11] R. Gharpurey, R.G. Meyer, "Modeling and Analysis of Substrate Coupling in Integrated Circuits", *IEEE J. Solid-State Circuits*, vol. 31, no. 3, pp. 344-353, May 1996.

[12] A.M. Niknejad, R. Gharpurey, R.G. Meyer, "Numerically Stable Green Function for Modeling and Analysis of Substrate Coupling in Integrated Circuits", *IEEE Trans. on Computer-Aided Design*, vol.17, no.4, pp.305-315, April 1998.

[13] T. Smedes, N.P. van der Meijs, A.J. van Genderen, "Extraction of Circuit Models for Substrate Cross-talk", *Proc. IEEE International Conference on CAD*, pp. 199-206, 1995.

[14] T. Smedes, N.P. van der Meijs, A.J. van Genderen, "Boundary Element Methods for 3D Capacitance and Substrate Resistance Calculations in Inhomogeneous Media in a VLSI Layout Verification Package", *Advances in Engineering Software*, vol. 20, no. 1, pp. 19-27, 1994.

[15] N.K. Verghese, D.J. Allstot, M.A. Wolfe, "Verification Techniques for Substrate Coupling and Their Application to Mixed-Signal IC Design", *IEEE J. Solid-State Circuits*, vol. 31, no. 3, pp. 354-365, March 1996.

[16] N.K. Verghese, D.J. Allstot, "SUBTRACT: A Program for the Efficient Evaluation of Substrate Parasitics in Integrated Circuits", *Proc. IEEE International Conference on CAD*, pp. 194-198, 1995.

[17] *SPACE: An Accurate and Efficient Layout-to-Circuit Extractor*, software and information available at http://dutentb.et.tudelft.nl/research/space.html.

[18] A.J. van Genderen, N.P. van der Meijs, T. Smedes, "Fast Computation of Substrate Resistances in Large Circuits", *Proc. IEEE European Design & Test Conference*, pp. 560-565, 1996.

[19] P. Miliozzi, L. Carloni, E. Charbon, A. Sangiovanni-Vincentelli, "SubWave: a Methodology for Modeling Digital Substrate Noise in Mixed-Signal ICs", *Proc. IEEE Custom Integrated Circuits Conference*, pp. 385-388, 1996.

[20] I.W. Wemple, A.T. Yang, "Integrated Circuit Substrate Coupling Models Based on Voronoi Tesselation", *IEEE Trans. on Computer-Aided Design*, vol.14, no.12, pp. 1459-1469, December 1995.

[21] K.J. Kerns, I.L. Wemple, A.T. Yang, "Stable and Efficient Reduction of Substrate Model Networks Using Congruence Transforms", *Proc. IEEE Design Automation Conference*, pp. 207-214, 1995.

[22] M. Pfost, H.M. Rein, T. Holzwarth, "Modeling Substrate Effects in the Design of High-Speed Si-Bipolar IC's", *IEEE J. Solid-State Circuits*, vol. 31, no. 10, pp. 1493-1501, October 1996.

[23] M. Pfost, H.M.Rein, "Modeling and Measurement of Substrate Coupling in Si-Bipolar IC's up to 40 GHz", *IEEE J. Solid-State Circuits*, vol. 33, no. 4, pp. 582-591, April 1998.

[24] R. Singh, S. Sali, "Efficient Modelling of Substrate Noise and Coupling in Mixed-Signal SPICE Designs", *Electronics Letters*, vol. 33, no. 7, pp. 590-592, 27th March 1997.

[25] L. Piessens, W.B. Vandervorst, H.E. Maes, "Incorporation of Resistivity Dependent Contact Radius in an Accurate Integration Algorithm for Spreading Resistance Calculations", *J. Electrochemical Society*, vol. 130, no. 2, p. 468, 1983.

[26] L. Deferm, C. Claeys, G.J. Declerck, "Two and Tree-Dimensional Calculation of Substrate Resistance", *IEEE Trans. Electron Devices*, vol. 35, no. 3, pp. 339-352, March 1988.

[27] D.K. Su, M.J. Loinaz, S. Masui, B.A. Wooley, "Experimental Results and Modeling Techniques for Substrate Noise in Mixed-Signal Integrated Circuits", *IEEE J. Solid-State Circuits*, vol. 28, no.4, pp. 420-430, April 1993.

[28] K. Joardar, "A Simple Approach to Modeling Cross-Talk in Integrated Circuits", *IEEE J. Solid-State Circuits*, vol. 29, no. 10, pp. 1212-1219, October 1994.

[29] N.K. Verghese, D.J. Allstot, "Computer-Aided Design Considerations for Mixed-Signal Coupling in RF Integrated Circuits", *J. Solid-State Circuits*, vol. 33, no. 3, pp. 314-323, March 1998.

4 Substrate biasing and noise coupling

In this chapter an analysis of different alternatives for substrate biasing is presented. In previous chapters we analyzed the characteristics of noise propagation and coupling to and from devices were analyzed. Now we will focus on the noise propagated through power-supply lines. Noisy supply lines may couple disturbances through the substrate contacts, and noise at the substrate may be coupled to analog supply lines, too. A simulation analysis is here presented, from which the advantages and drawbacks of bias approaches, number of substrate contacts, or the effect of pads, are discussed. The chapter also contains an evaluation of the probability of latch-up in some of the biasing situations proposed, triggered either by noise at supply lines or by substrate noise.

4.1 Introduction

In a mixed-signal circuit independent supply lines for the analog and digital circuitry are commonly used. This is a way to avoid noise coupling from lines feeding the digital circuitry to the analog signals. Nevertheless, taking into account noise propagation through the substrate, a number of doubts appear when designing substrate biasing. Is it better to bias using analog or digital *GND* lines? If the digital ground is used, it is strongly contaminated with switching noise, and contacts will introduce this noise to the substrate. If the analog line is used, substrate noise is acquired through the contacts and affects the circuitry connected to it. One choice is to use lines dedicated exclusively for biasing purposes, but this implies using a number of contacts which may not be available, or using larger packages with the consequent cost increase. The number of questions to respond to is much larger, though. Is a guard ring much effective? Where should be the ring connected? Given a biasing strategy, is it better to use many or just a few contacts? May the pad ring affect the biasing strategy? According to what was seen in chapter 2, all these questions must be asked both for P+ and P- wafers, because their characteristics may lead to radically different answers in every case.

In this chapter we will discuss these questions, thus completing the analysis and recommendations presented in [1], [2].

We have analyzed a low complexity mixed-signal circuit as a reference case from which start discussion. We assumed that the circuit is implemented in a 1.2 μm N-well CMOS technology, 2 metal layers. The digital part consists of a combinational tree driving a number of noisy inverters. These noisy elements are organized in 16 blocs of 16 inverters each. Every inverter is loaded with the gate capacitance corresponding to another identical inverter, which provides transition times of the noisy drivers of 0.8 ns, for both the trailing and falling edges.

A simple amplifier with short-circuited input and output is chosen as noise-sensing analog element, with DC voltage at the sensing node thus lying around $V_{dd}/2$, 2.5 V in our case. The sensor transistor sizes are 50/1.2 for the PMOS and 20/1.2 for the NMOS, and the output node is loaded with the capacitances of another identical element. Independent power-supply lines for the digital and the analog section are assumed, and digital pads have their own supply line. Communication to the outside is done through 30 digital pads and 10 analog ones.

A layout of the circuit is generated in order to extract parasitic capacitances, taking care to duplicate supply lines on the circuitry so as to allow substrate biasing with a line different to that powering the transistors. Supply lines in the analog section are larger than strictly necessary, in order to give them realistic sizes, approximately a forth of those of the digital section. Capacitance values between nodes are obtained with an extractor. The only capacitances that have shown to influence simulation results are the capacitances between supply lines in the same section, and those between the lines and the substrate, as shown in Figure 4.1. The different supply lines are connected to ideal external voltage sources through the equivalent circuit shown in Figure 4.2, which models the package parasitic effects [3]. A full model would also include the mutual parasitic elements between pins [4], [5], but considering these effects would imply assuming a particular pin assignment. The most significant element in Figure 4.2 is the inductance, for which a 10 nH value is taken, typical of a large number of packages [5].

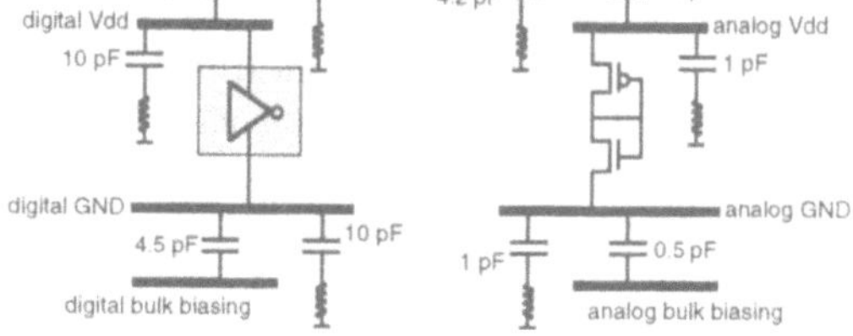

Figure 4.1

Capacitances significant in noise evaluation, with the values obtained from the extraction of the reference circuit. Capacitances between well biasing lines and the substrate are due to the junctions of the wells.

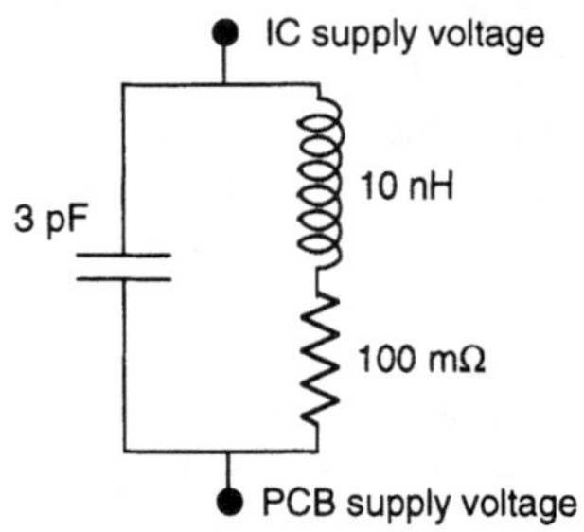

Figure 4.2

Equivalent circuit chosen to model the effects of package pins [3].

The extraction of substrate resistances is performed with the device simulator MEDICI. A certain regularity in the layout will be assumed to simplify resistance extraction. Every digital gate has a contact close to it, thus resistance extraction can be obtained from a 2-D simulation of a vertical slice like that shown in Figure 4.3. The substrate model chosen is then a full model according to the terminology used in chapter 3, because we consider all the resistances between the ports shown in the slice of figure 4.3. Every noisy inverter disturbing the sensor will have associated to it a resistance net like that of the figure. With another extraction of a slice perpendicular to the one represented in the figure, the resistances between two contacts, or between two transistors in the digital section may be obtained. Nevertheless, given that inverters toggle synchronously, and given that contacts are biased by the same line, those resistances are not necessary. For the sake of greater simplification, only the coupling net between the sensor and the four closest noisy blocks will be considered. Apart from the resistances between ports, the model includes the resistance between every contact or transistor and the substrate node. In P- wafers this has no meaning unless a conducting backside plane is assumed, which is common in most of packages.

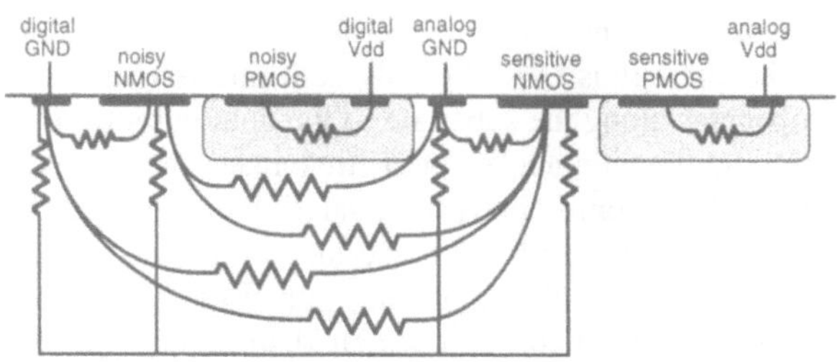

Figure 4.3

Diagram showing the resistances extracted between the ports in a vertical slice of the circuit.

Throughout this chapter and particularly in the figures, we will use a determined terminology to identify the different biasing strategies. So, when an *A/B* biasing is mentioned, the first letter indicates the line to which contacts in the digital section are connected, while the second letter indicates the lines that bias contacts in the analog section. If the layout contains a guard ring enclosing the analog section the term *A/B/C* biasing will be used, where the central letter identifies the line that biases the guard ring, and the first and the third ones identify the lines biasing the digital and the analog sections, respectively. Identification of the different lines is done by means of letter *D*, for the line feeding the digital circuitry; *A*, when the line feeds the analog section; and *E*, when it is a line that does not provide current to any transistor, but is only used to bias the substrate or the wells.

4.2 Biasing P- substrates

4.2.1. Coupling mechanisms and biasing strategy

The reference circuit presented in the former section is simulated with transient analysis, the digital part excited by a 20 MHz clock signal. Only 4 out of 16 noisy blocks are enabled, and results presented in this chapter are obtained in these conditions unless stated otherwise. Contacts in digital and analog sections are connected to combinations of the three supply lines available, which are the one for digital circuitry, one for analog circuitry, and a line exclusively used for biasing purposes, not feeding any circuit.

In order to make this easier to understand, we start by analyzing the circuit without the pads and without guard rings. In Figure 4.4 the results obtained for the strategies considered are shown. The large amount of noise when connecting all contacts to the analog supply stands out. With this A/A strategy, the noise level in the analog supplies is about five times larger than in the other situations. Noise is acquired by contacts near the noisy gates, and transmitted to the sensor through the supply line. On the other hand, when the strategy is D/A, the only path through which noise can affect the analog supply propagates along the substrate to the analog section, and then couples to the supply lines through the contacts or parasitic capacitances. Given the resistivity of the substrate, noise will be significantly attenuated. Note that this last strategy –D/A– is better than D/D or even than E/E. In these last two cases, even though the analog supplies are not connected to the substrate, substrate bias lines shortcircuit both domains thus providing a path through which disturbances at the digital part may reach the analog part and thus affect devices. Strategies D/A and E/A minimize body effect but continue to involve the problem of a certain amount of noise transmitted through the supplies, while the best strategy that solves all of these problems is D/E. Lastly, Figure 4.4 also shows the noise level assuming ideal supply pins (no SSN), i.e.

noise directly transmitted from the switching transistors to the quiet transistors. In short, D/A biasing achieves a coupling low enough without any need for dedicated lines, whenever a suitable distance between coupled circuits is maintained in order to allow noise attenuation. If technological characteristics or layout restrictions make a significant amount of noise unavoidable in the analog part, it is advisable to bias this part with dedicated lines, thus avoiding noise transmission to the analog supplies.

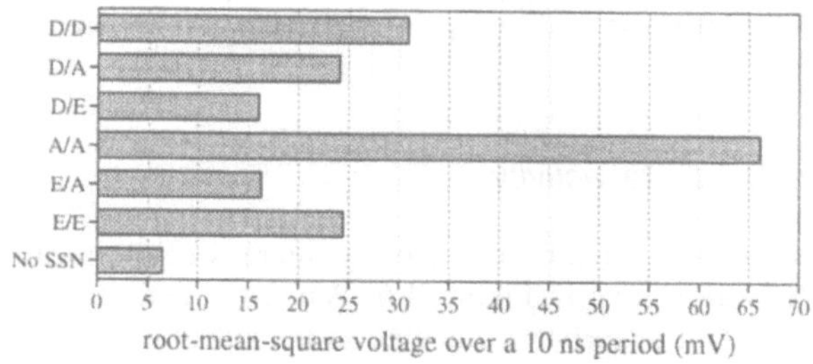

Figure 4.4

Noise levels according to the biasing strategy. P- substrate, no pads.

As we have just seen, the main source of trouble is the noise coupled to the supply lines through the contacts. It might be thought that the amount of transmitted noise could be reduced if the number of substrate contacts were minimized. In order to verify this assumption the previous analysis is repeated, but this time multiplying the values of the resistances between contacts and any other node by 10, which is equivalent to reducing the number of contacts by a similar factor. The results obtained are shown in Figure 4.5, and it can be seen that the effect is not the one expected, but quite the opposite, indeed. By observing the noise level when ideal supplies (no SSN) are available, the main reason of the general noise increase is seen to be the larger

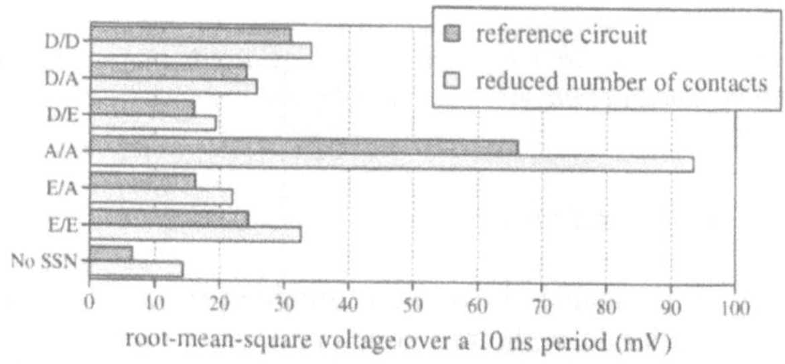

Figure 4.5

Noise levels according to the biasing strategy, reducing the total number of contacts. Comparison with results in Figure 4.4.

direct coupling between transistors. The reason of this increased coupling is that the disturbance has now a higher impedance return path to ground, while keeping the value of the impedance of the path to other devices. On the other hand, it is observed that reducing the number of contacts increases the noise in the supply lines, especially the ones biasing the digital section. The consequence of reducing the number of contacts will then depend on which is the dominant coupling mechanism: if noise is mainly coupled through the supply lines, increasing the number of contacts will increase noise, while if noise is directly coupled between devices, more contacts will reduce noise.

4.2.2. The effect of digital activity

The noise levels obtained for different activity degrees of the switching section, 4 or 16 blocks enabled, are shown in Figure 4.6. When the activity of the digital section increases, the amount of disturbance injected to the substrate is larger, which favors a generalized increase in noise at the sensor, and an increase in noise in the line biasing the digital section. Furthermore, the noise at the digital supply lines also increases very significantly –it doubled, in this case–, so contacts biased by this line introduce even more noise, like in D/A or D/E biasing.

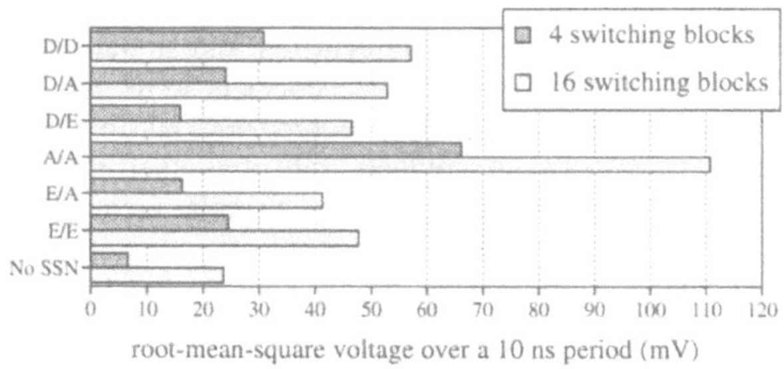

Figure 4.6

Noise levels according to the biasing strategy used, increasing the activity of the digital section. Comparison with results in Figure 4.4.

4.2.3. The effect of the pad ring

The simulation results presented above were obtained for a reference circuit without the pad ring, in order to simplify the problem and make it more easily understandable. Now we shall determine whether this simplification has led to an error in the results, and in any case find out the influence of the chip periphery on the noise received, depending on the biasing strategy. The ring is formed by 30 digital signal pads, 10 analog signal pads, and 8 pads corresponding to the digital core, digital periphery, analog and dedicated supplies. Whatever the biasing strategy in the circuit core is, the contacts in the pads are connected to the corresponding supply lines, digital if digital pads, analog if analog pads.

The results obtained at the output node are shown in Figure 4.7, and it can be seen that the effect of pads is in general not very important, limited to a noise damping in the supply lines, which implies a reduction of the amount of noise that reaches the sensor. The reason for this damping can be found in the increase of the capacitance between power lines and the substrate, and in the effect of substrate contacts connected to ground lines distant from the noisy section. The improvement is greater when the noise at the sensor output is mainly due to its supply, A/A strategy. It has been proved that increasing the number of the different kinds of pads does not lead to significant variations in the results.

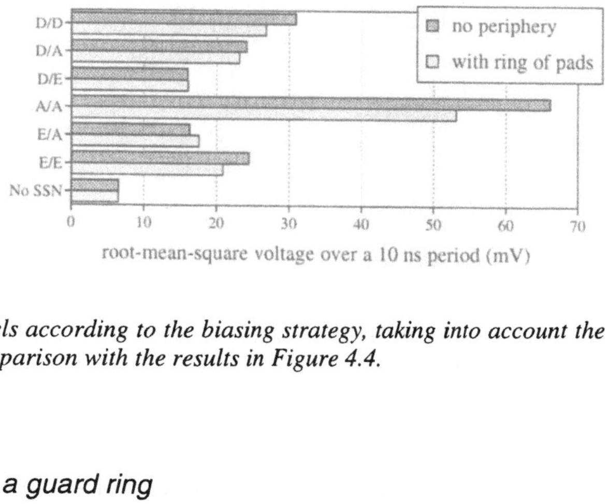

Figure 4.7

Noise levels according to the biasing strategy, taking into account the effect of the pad ring. Comparison with the results in Figure 4.4.

4.2.4. Biasing a guard ring

In chapter 2 it was seen that, particularly in P- substrates, a significant noise reduction could be achieved with guard rings. That analysis was done under ideal conditions, and the different biasing alternatives were not taken into account. Now simulations are done including a guard ring enclosing the analog circuitry, only for the best biasing strategies according to the results in Figure 4.4. If Figure 4.3 is observed, it can be seen how, in the layout taken as a reference, substrate contacts in the analog section are placed between noisy gates and the sensor, thus acting as a guard ring themselves. Consequently, including a new ring will not have as significant an effect as the one obtained when there is nothing in the way between noisy and sensitive transistors. In any event, the point of interest is finding out the advantages and disadvantages of biasing the ring with the different supply lines available.

Focusing on the two first options shown in Figure 4.8, D/E/A is more advantageous than D/A/A. The main reason for this is that D/E/A biases the substrate in the analog section with two different lines, thus reducing the equivalent impedance to the external ground. When laying a ring and biasing it optimally (D/E/A), it will be important to prevent the analog supplies from acquiring noise despite the presence of the ring. That is to say, a ring enclosing a sensitive circuit will be useless if there are analog contacts allowing a path through which noise affects the circuit, because of the number of these contacts or because of their proximity to noisy sources.

If a ring is to be added to an E/A biasing, the E/E/A option is clearly worse than E/A/A. This is again because in the first case the same line is used to bias the digital section and the ring, providing a low impedance path for noise to reach the analog section. This is the reason why results of strategies like D/D/A or D/D/E are not presented in Figure 4.8 either, these results being similar to those of D/D biasing.

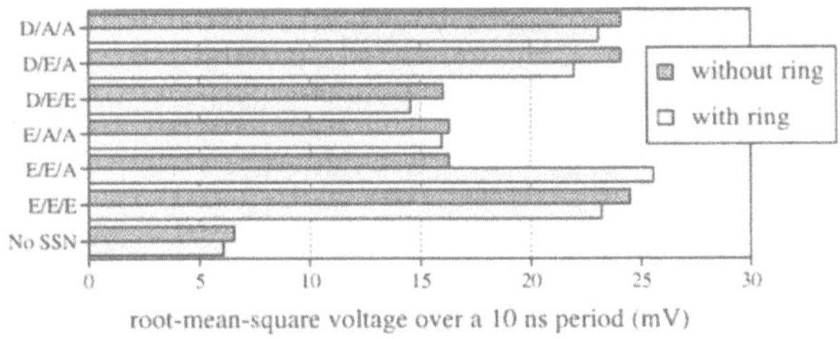

Figure 4.8

Noise levels according to the biasing strategy, including a guard ring. Comparison with results in Figure 4.4.

4.2.5. The effect of the inductances of power supply terminals

The results presented up to this point assumed 10 nH inductances in all terminals. Even though it is a typical value in many packages, in practice values from a few nanoHenries, such as in SOIC packages, to tens of nH (large QFP) can be found [6], [7]. Moreover, inductance value varies very significantly both in terms of the relative position of the pin in the package, and in terms of the frequency. It is thus advisable to find out whether the differences between biasing strategies are kept when the inductance is lowered. The simulations presented in section 4.2.1 are repeated now assuming 1 nH inductances, and results are shown in Figure 4.9. It can be seen that for very low inductances, noise levels are made uniform, thus disappearing the large differences that existed between the strategies. If noise at the different nodes is observed, it can be seen that a drastic noise reduction in the supply lines is produced (they now have ten times less impedance to ground), and thus noise at the sensor output is mainly due to direct coupling between transistors through the substrate. When the inductance value increases, the relative importance of noise coupled through the supplies is larger, and the biasing strategy becomes more important. Noise at the sensor depending on the inductance of the package pins is represented in Figure 4.10, D/A biasing, and it can be seen that the larger its value, the larger the noise coupled. This shows that the problem of coupling between digital and analog circuitry not only has to be taken into account during chip design, but also when making decisions connected with the package. Choosing packages with low parasitic inductances will be important, and then connecting the supply pads to the low inductance pins, especially analog supplies and those biasing the substrate in the analog section. Multiple pin assignment to a single line is an efficient technique for reducing effective inductances, that was already proposed in [8], and algorithms for choosing pins so as to reduce switching noise can be found in [9], [4]. A study of supply lines pin assignment in order to minimize noise can also be found in [6].

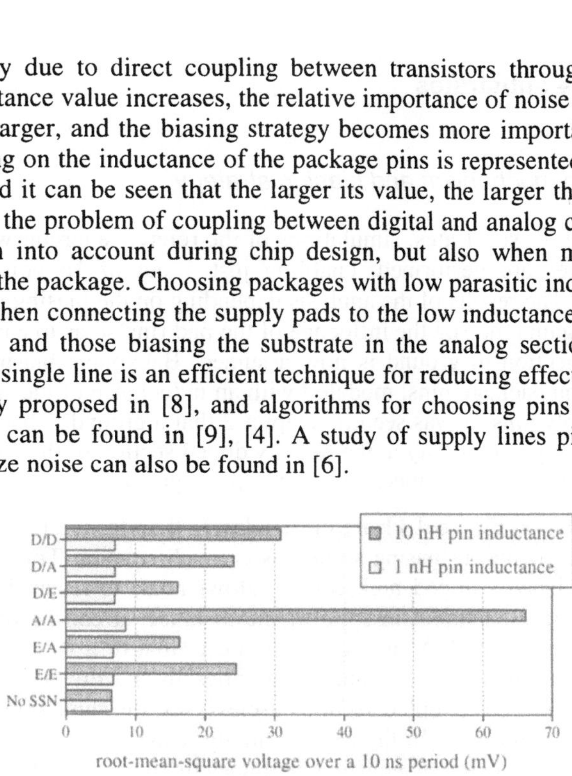

Figure 4.9

Noise levels according to the biasing strategy used, with supply pin inductances reduced to 1 nH. Comparison with results from Figure 4.4.

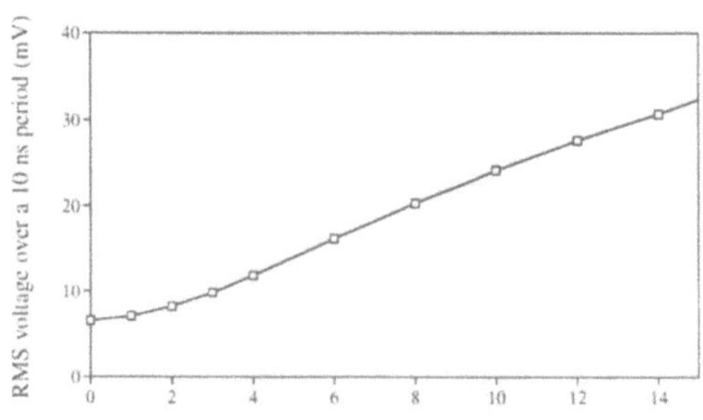

Figure 4.10

Substrate coupled noise dependence on the power supply pin inductance. P- substrate, D/A biasing.

4.3 Biasing P+ substrates

4.3.1. Coupling mechanism and biasing strategy

As was done with P- substrates, simulations of the reference circuit when biased with different strategies are performed, enabling just 4 out of 16 noisy blocks unless otherwise stated. The results of the analysis depending on the biasing policy are shown in Figure 4.11, again ignoring the influence of the pad ring so as to ease interpretation. A backside connection to ground is now assumed. Backplane biasing will be made through one of the package pins, modeled with an inductance of the same value as the other power supply pins. It was seen in chapter 2 that a grounded backplane offered a large noise-reducing potentiality at the heavily doped section of the substrate, which is the part through which disturbances are mainly propagated.

Noise levels shown in Figure 4.11 are clearly larger than those obtained in P- wafers, and the difference between biasing strategies is much smaller. The reason is the low resistivity of the heavily doped bulk, which allows noise to reach the analog section with no significant attenuation, in spite of the distance. It can be observed how the noise level obtained in an A/A strategy is similar to that of a P- wafer. In the former case noise was introduced to the analog supplies in the digital section, making a distance increase between coupled sections useless, just the same effect as is produced by a heavily doped substrate. In the case of highly resistive substrate, it was said that separation favored insulation when using strategies like D/A or E/A, while now analog contacts acquire substrate noise regardless of their placement. Moreover, the disturbance is coupled to the analog supply not only through the contacts, but also through the capacitances of the lines to the substrate. Hence, strategies like D/E which produced low noise levels in P- substrates, both at the sensor and at the analog supplies, now present high noise levels at all nodes.

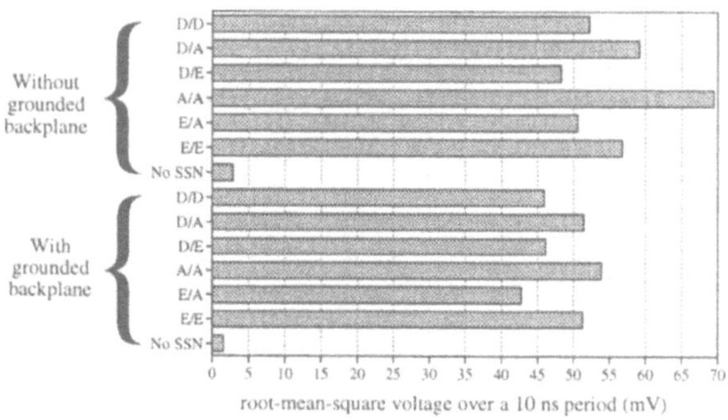

Figure 4.11

Noise levels according to the biasing strategy. P+ substrate, no pads.

It can also be observed that in the limit case of ideal supply lines –no SSN–, noise measured at the sensor in a P+ wafer is amazingly about three times smaller than that of a P- wafer. The large number of contacts in a real size circuit largely reduces the impedance between the bulk inside and the supply lines, so if they were ideal supplies bulk would be almost perfectly biased. This is illustrated in Figure 4.12.a: from a noise propagation viewpoint, disturbance finds a very low impedance return path to ground through the supply lines, much better than the return path through the analog devices. From the noise levels viewpoint, the situation may be modeled as a voltage divider formed by the resistance between noise sources and the substrate, the impedance between the substrate and supply lines, and the impedance of the pin between these lines and the external ground. If the last two impedances are much smaller than the first one, voltage levels at the bulk will be near zero and no noise will propagate to the analog devices. On the contrary, if pin impedances are larger than the equivalent resistance under the switching devices, noise level at both the supplies and the substrate will be high, regardless of what biasing strategy is used. In other words, the return path through the power supplies is then comparable to the one through the analog circuitry.

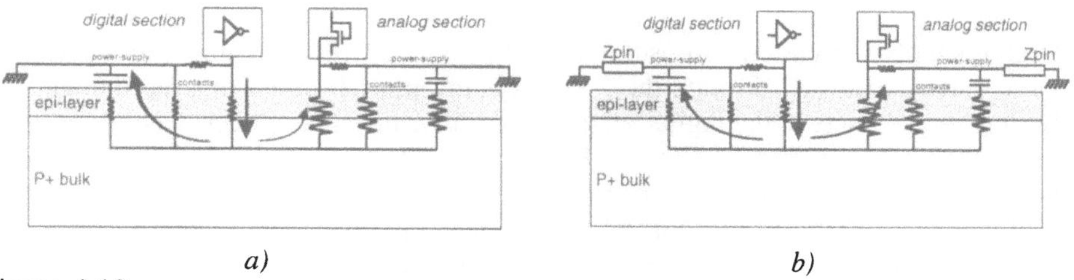

Figure 4.12

Illustration showing noise propagation in an IC manufactured on a P+ wafer. a) In case power supply pins were ideal. b) When the value of the impedance of the supply pins is high compared with the impedance between noise sources and the substrate.

Noise propagation to the analog part is consequently almost unavoidable in this type of substrate, given the usual values of package parasitics. Not much can be done with the biasing strategy, only attempting to maximize the impedance between analog supplies and the substrate (avoiding using them to bias the substrate, minimizing their width and dimensions), and also to maximize the number of pins connected to supply lines (to reduce their effective impedance).

Going back to the results in Figure 4.11, adding a backplane does not provide a large reduction in noise levels, or at least the added advantage is much smaller than expected from the analysis in chapter 2. The reason for this is similar to the one explained, which is illustrated in Figure 4.13. Substrate can be modeled in a simplistic way as a voltage divider formed by the impedance between noise sources and the substrate, across the epi-layer, and the impedance of the pin connected to the backside contact. If the switching part has some complexity (thousands of gates), the first of the impedances is lower than the second one, and thus bulk is not properly biased.

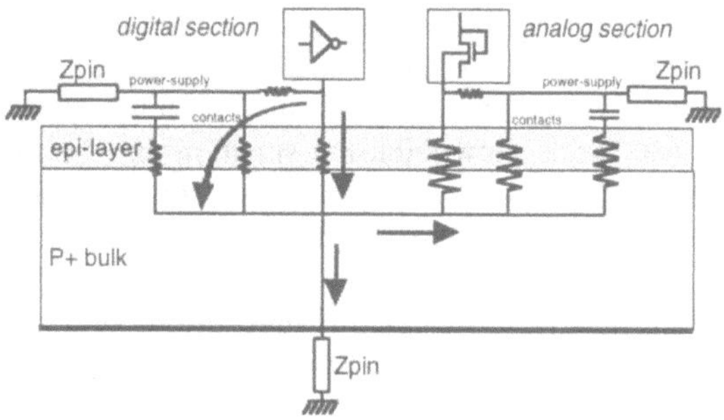

Figure 4.13

Illustration showing noise propagation in an IC manufactured on a P+ wafer, with a backside contact.

Digital supply lines are usually strongly contaminated with switching noise, given the large currents flowing through the corresponding pins. It could thus be thought that biasing the contacts in the digital section with those lines introduces even more noise, given their number. If substrate was biased with a backplane, surface contacts could be left out and then the amount of noise be reduced. In order to verify this assumption, simulations of the different biasing approaches are made under those conditions, removing substrate contacts in the digital section. The results are shown in Figure 4.14, and it can be seen that it is only when a single line biases the whole IC –

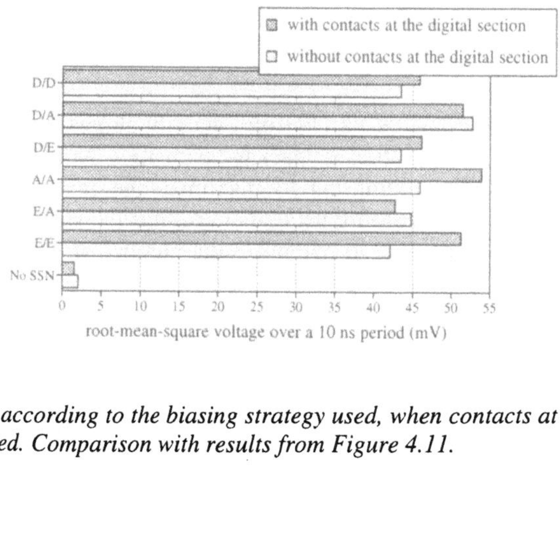

Figure 4.14

Noise levels according to the biasing strategy used, when contacts at the digital section are eliminated. Comparison with results from Figure 4.11.

A/A, E/E–, that a significant noise reduction is achieved. In the remaining situations noise levels remain similar, which means that in those cases the majority of the disturbance was not injected by the contacts, but directly by the devices through their junctions.

4.3.2. The effect of the pad ring

As was done in the former section, the extraction of the parasitic elements of a ring formed by 48 pads, with the same distribution and characteristics detailed in section 4.2.3, is added to the reference circuit. The results obtained are shown in Figure 4.15, and it can be seen that, unlike what happened in P- substrates, the ring of pads totally changes the results seen so far. Noise levels are consequently reduced and are made even more uniform, now absolutely independent of the biasing strategy.

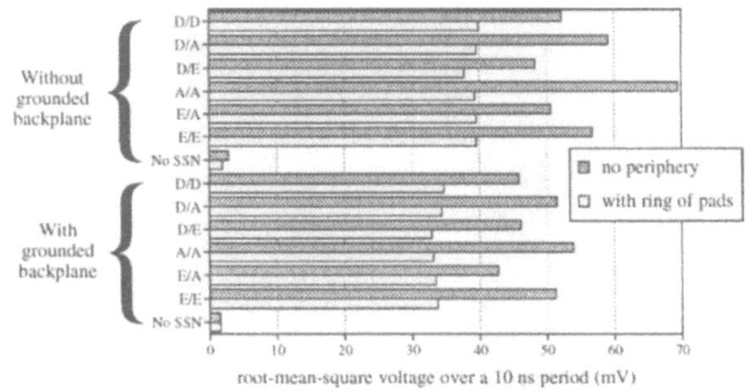

Figure 4.15

Noise levels according to the biasing strategy used, taking into account the effect of the ring of pads. Comparison with the results from Figure 4.11.

The reduction of noise levels due to the presence of the pads was already observed in P- substrates, although not as much as now, and it was due to the damping of the oscillations in the supply lines, because the ring involves large capacitances between these lines and the substrate. It is common to find supply lines about one hundred microns wide linking the pads, forming a large ring enclosing all the circuitry in the IC. Nevertheless the most significant result is the uniformity of noise levels, which means that the effect of capacitances and contacts in the ring of pads is much larger than that of the circuit core, and this masks the effect of the different biasing strategies. For example in Figure 4.16 the effect of eliminating the contacts in the digital section may be observed, and it can be seen how this effect can now be totally ignored, while in Figure 4.14 (without the pads) the effect was significant in some biasing approaches like E/E or A/A.

If noise waveforms are observed, they look like damped oscillations of similar frequency and amplitude in all cases. Disturbance at the digital supply is injected to the substrate through contacts and parasitic capacitances in the pads, regardless of the biasing strategy at the circuit core, and it is collected by the analog pads, thus affecting the analog circuitry through its supply. In P- substrates, the disturbance hardly reaches the analog pads because of its attenuation, and hence there the noise received depends on the biasing characteristics of the core. A number of simulations were made, altering the number of pads, and the characteristics of the noise received were indeed modified. If the number of switching blocks is increased, the noise level increases too, although it keeps being practically independent of the biasing strategy.

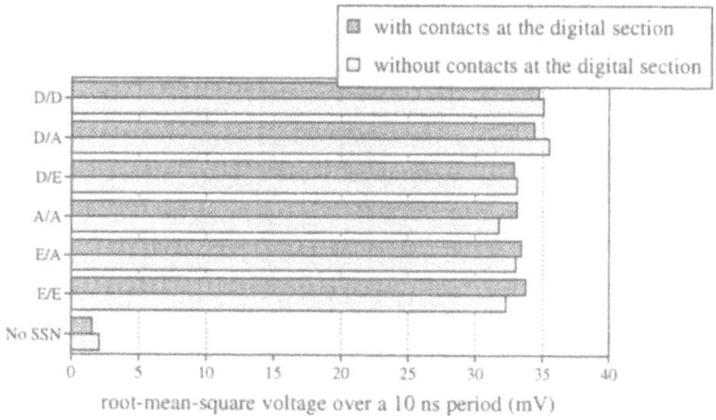

Figure 4.16

Noise levels according to the biasing strategy used, when contacts at the digital section are eliminated. P+ substrate with a grounded backplane, now taking into account the ring of pads. Comparison with results in Figure 4.15.

4.3.3. The effect of inductances of power-supply terminals

In order to complete the study, the effect of having low pin inductances was determined. Figure 4.17 compares the noise levels obtained with 10 nH and with 1 nH inductances, including that of the pin connected to the backside contact. Simulations were made including the ring of pads, given its influence. Two effects stand out when observing the results. On one hand, the difference between biasing strategies is surprisingly larger with 1 nH inductances than with 10 nH. If inductance values are low enough, noise at the supply lines –digital ones in particular– diminishes, and then the main contribution to substrate noise is not the disturbance introduced by the pads, but the noise injected by the devices themselves. So in this case the connection of the contacts is significant.

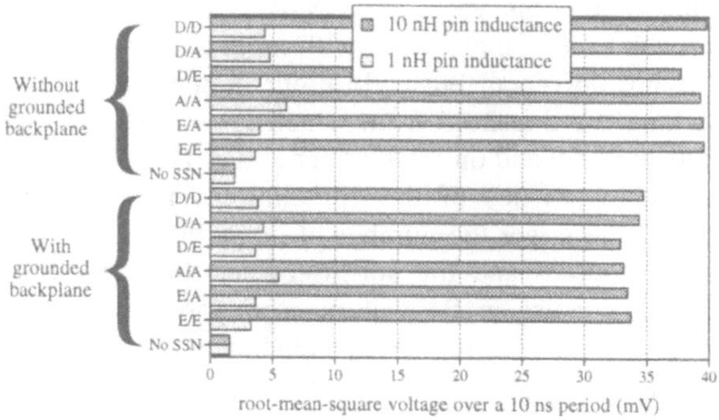

Figure 4.17

Noise levels according to the biasing strategy used, supply pin inductances reduced to 1 nH. P+ substrate. Comparison with the results from Figure 4.15.

The second effect that stands out is the large noise reduction obtained when the value of the inductances is reduced. This reduction is even more evident when comparing Figures 4.10 and 4.17: with 10 nH, P+ substrates are noisier than P- ones, while with 1 nH it is quite the opposite case. Once again the explanation can be found in point 4.3.1 when talking about noise propagation and the role of pin impedances. The characteristics of P+ substrates make the substrate perfectly biased and free of noise if the pins are ideal or almost ideal –low inductances. On the contrary, if the pin impedances are large in relation to the impedances between circuitry and substrate, noise is not eliminated. Thus, a threshold inductance value will exist, under which it will be better to use P+ wafers than P- ones. Noise dependence on inductance is represented in Figure 4.18, for both kind of wafers and D/A biasing; in the P+ case a grounded backplane is assumed. At an inductance value of 2 nH the noise level in

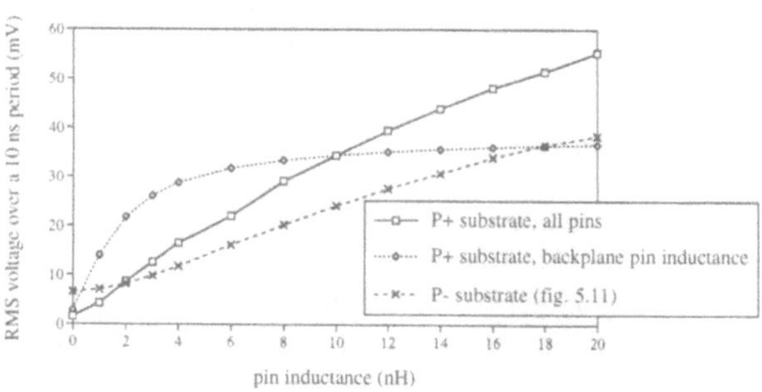

Figure 4.18

Substrate coupled noise dependence on all the pin inductances, and only on the backside contact pin inductance. P+ substrate with grounded backside contact, D/A biasing. Comparison with the results from Figure 4.11.

both wafers is the same. This value is very low, and most packages have larger inductances, but it can easily be achieved with multiple pin assignment. Nevertheless this 2 nH value is not universal, but a different inductance value can be found for every circuit, marking the threshold from which P+ wafers are worse than P- ones. This threshold value will depend on the number and dimensions of noisy nodes and on the technological characteristics of the epitaxial layer. In general, the larger the complexity of the IC, the larger the number of noisy nodes, the lower the impedance between these noise sources and the substrate, and thus the lower the inductance valueneeded to get low noise in P+ wafers. Remember that the circuit taken as a reference is a small one, so the threshold inductance value will normally be much smaller than 2 nH, usually too small to achieve in all supply pins. Given the characteristics of P+ wafers, one could consider obtaining this value only for the backside contact pin, which could be easier to achieve by multiple assignment. A third curve is shown in Figure 4.18, which shows the noise dependence on the inductance of the pin connected to the backside contact, while the other supply lines keep connections through 10 nH inductances. It can be seen that this proposal is not viable, because the threshold value is now even smaller. Noise at the supply lines is now kept, so the degree of ideality demanded of the contact is larger than when reducing all inductance values at the same time.

4.4 Latch-up implications

In previous sections it was seen that biasing strategies like E/A or D/E provide low substrate noise, particularly in lightly doped substrates. Nevertheless, those strategies imply biasing the substrate in the analog or the digital section with different lines from the ones feeding the circuitry. Even though DC values in both lines will be the same, noise levels are different and they might be large enough to induce latch-up problems. We will now briefly determine whether amplitude and frequency characteristics of noise at supply lines and substrate may be enough to induce latch-up.

4.4.1. The latch-up problem

Latch-up is a phenomenon inherent to CMOS processes, where two parasitic bipolar transistors appear between the substrate, the well, and the sources of a pair of transistors, as shown in Figure 4.19. These bipolar transistors are connected forming a SCR thyristor structure, and the conduction of one of the devices may thus give rise to positive feedback that ends in a stable state in which voltage at the emitter of the NPN lies around 4 V –holding voltage–, creating an MOS malfunction. This stable state can only be removed by turning off the power supply. The worst consequence of latch-up is that conducting bipolar transistors allow a current path between positive power and ground and, depending on the values of resistances R_s and R_w in Figure 4.19, the current may be large enough to produce circuit breakage, usually because of the destruction of power supply lines. A single *npnp* structure may be not sufficient to destroy the chip, but the disturbances led into the substrate may switch on other close *npnp* structures, thus extending the effect to the point of destruction. The values of resistances R_s and R_w depend on technology characteristics and on the placement of the contacts in the well. The closer the contacts to the devices, the lower the resistances and the lower the probability of triggering latch-up. The design rules on maximum distance between substrate contacts given by IC foundries are obtained in view of the risk of latching. In fact, P+ epitaxial substrates, which in this chapter have been proven to be inadvisable from the noise viewpoint, are used in order to minimize resistance R_s and thus avoid latch-up.

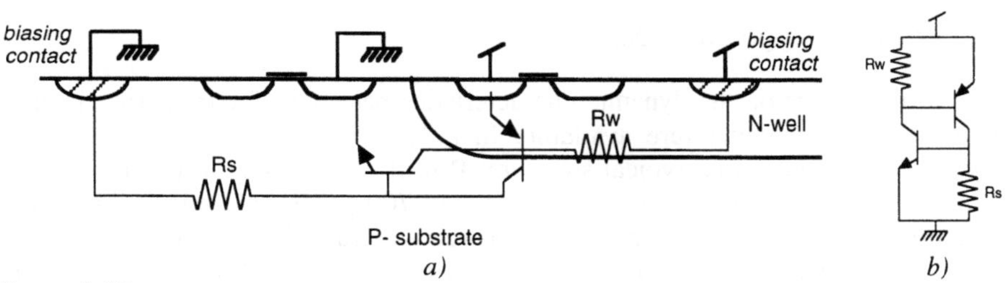

Figure 4.19

a) Description of the parasitic elements appearing in latch-up. b) Equivalent SCR circuit.

Among the causes that may trigger latch-up, we can find reflections or noise in the lines connected to the MOS sources, outputs connected to a powered down circuit, electrostatic discharge, displacement currents originated during a circuit power up, substrate currents due to leakage or hot carriers, or external radiation. The following three conditions must be met if one of the former causes is to produce latch-up [10]:

• Base-emitter junctions of both parasitic transistors must be turned on, which means V_{be} voltages in the order of 0.6-0.7 V.

• The product of current gains of both transistors ($\beta_{npn} \cdot \beta_{pnp}$) must be large enough. Expressions for the condition of sufficiency may be found in [11], [12].

• The power supply must provide enough current to maintain the phenomenon.

The third condition is not restrictive, unless working with current limiters, so latch-up prevention techniques are based on reducing β_{npn}, β_{pnp}, because of the second condition, or reducing resistances R_s and R_w. Typical values of current gain of the vertical PNP are between tens and hundreds, depending on doping and depth of the N-well implants, while gains for the lateral NPN are under 10, strongly depending on the distance between the N-well and the NMOS. Values of the resistance R_s may be a few KΩ if contacts are not far away from the transistors, and tens of KΩ for R_w.

Among the measures for preventing latch-up, we may find β_{npn} reduction by increasing the distance between the NMOS and the N-well, although this is not an attractive solution due to its non-sufficiency and area increase; β_{npn} reduction by laying buried layers under the wells; R_w reduction by increasing the number of contacts or surrounding the well with N+ rings; R_s reduction by increasing the number of contacts, placing P+ guard rings between the NMOS and the well, or using P+ wafers. A number of less common technological or layout techniques have also been proposed in the literature (see [10], for an example).

4.4.2. Latch-up triggered by substrate noise

Some partial studies on the dynamic characteristics necessary for triggering latch-up may be found in the literature available. In [13] a MEDICI analysis is made by applying a 2.5 V pulse to a typical structure, P-well on an N+ bulk process, finding that latch-up appears when pulses are longer than 700 ps. In [14] a two-dimensional device simulator is also used to analyze an N-well structure, 2 μm process, applying a negative voltage step 3 ns long to the emitter of the NPN lateral transistor. It is seen that the NPN transistor turns on at 2.4 ns, while the PNP does so at 2.8 ns. Lastly, [15] contains a description of a simulation tool with which a P-well structure is analyzed, in a 0.5 μm N+ process powered at 3 V, where an 18 ns long step of -2 V is applied to the emitter of the vertical NPN. As a result of this pulse, the NPN transistor switches on at 410 ps, while the lateral PNP does so at 12.5 ns. These analyses are highly specific and do not give an idea of the relationship between pulse amplitude and duration necessary for starting latch-up. Moreover, turning on the parasitic transistors does not guarantee feedback if the exciting pulse disappears immediately.

Taking the equivalent circuit in Figure 4.19.b and using parasitic transistor models with mixed-signal 1.2 μm technology, we determined the amplitude and duration conditions that a noise pulse propagating through the substrate must accomplish to turn the parasitic transistors on and start positive feedback. It is assumed that noise is produced by a close switching device, so the pulse affects the NPN transistor base through a resistive path. A diagram of the circuit simulated is shown in Figure 4.20. Power supply lines are considered to be free of noise.

A β_{pnp} value of 200 is taken for the vertical PNP transistor, while β_{npn}=10 for the lateral NPN one. Unfavorable values of R_s=10 KΩ and R_w=20 KΩ are chosen for the resistances to the contacts. As regards resistance R_c between the noise source and the NMOS transistor, the especially unfavorable value of 1KΩ is taken. If the substrate is properly biased, R_s and R_c should be similar, or the first smaller than the second. Simulations of the circuit are made, with different noise pulse amplitudes, and the minimum pulse width to trigger the feedback is determined. The results are shown in Table 4.I.

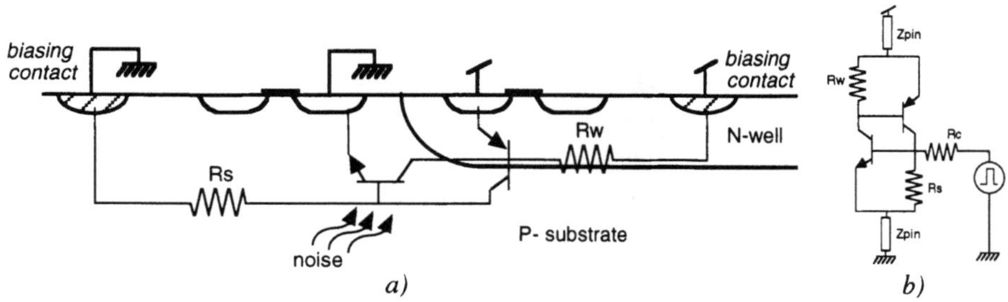

Figure 4.20

a) Latch-up triggered by substrate noise. b) Equivalent circuit simulated.

Table 4.I.

Minimum width of a noise pulse in the substrate required to trigger latch-up, as a function of its amplitude.

Pulse amplitude (V)	Minimum width (ns)
2	4.5
1.5	6
1	9
0.75	19

The smaller the pulse amplitude, the lower the duration necessary for sustaining latch-up, and the minimum width depends on the difference between the noise amplitude and V_{be}. A minimum duration of 4.5 ns is necessary for a 2 V amplitude, which is a demanding value insofar as noise pulses are no longer than a few ns and their amplitudes are under 2 V. In fact, if noise is injected into the substrate through switching drain junctions –leaving aside noise from supply lines–, the noise amplitude will be no larger than approximately 0.75 V, because this voltage implies turning on the source-bulk junction, thus preventing larger noise. So, assuming a maximum noise voltage of 0.75 V in the substrate, pulses maintaining this voltage for tens of nanoseconds will be necessary, a situation which can be dismissed according to our previous analysis. Figure 4.21 shows the currents through the bipolar transistors when the substrate is excited with a 0.75 V noise pulse, and it can be seen that feedback is produced if the noise width is equal to 19 ns, while when pulses are shorter the current returns to zero once the pulse ends.

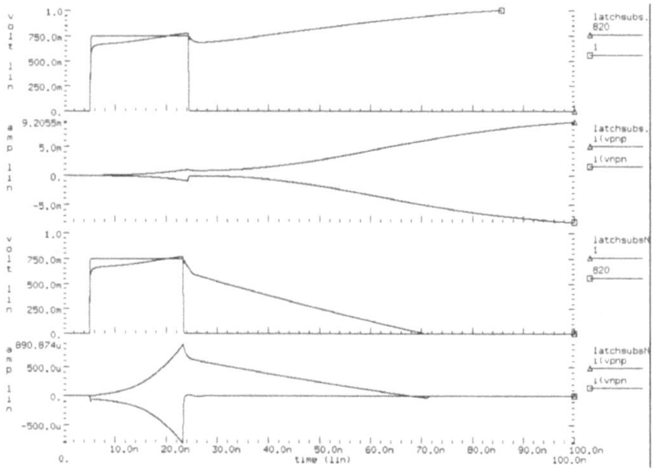

Figure 4.21

Waveforms of the exciting pulse and NPN transistor base voltage, and of the currents through the transistors, when noise amplitude is 0.75, and its duration is 19 ns –up– and 18 ns –down–.

4.4.3. Latch-up triggered by noise at the supply lines

After establishing that a substrate noise pulse can trigger latch-up, a check is also made to see whether noise at the supply lines can cause this effect, particularly when biasing the substrate with a line different to the one powering the MOS transistors. The reference circuit is the same as for the simulations in section 4.2, which contained 16 noisy blocks. This circuit is connected to a SCR structure as shown in Figure 4.22, i.e., keeping substrate contacts free of noise, while the sources of both transistors are

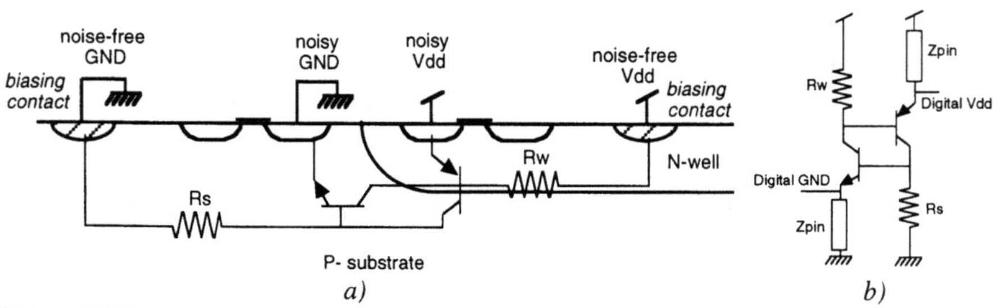

Figure 4.22

a) Supply lines connection to verify latch-up triggered by noise at these lines. b) Equivalent circuit simulated (supply lines connected to the reference circuit in section 4.2)

connected to the lines powering the switching blocks. Specifically, we want to know the activity degree and pin inductance values required to trigger latch-up. Note that these two characteristics are the ones determining the noise level and frequency at power-supply lines.

The results of the analysis are shown in Table 4.II, specifically the minimum number of noisy switching blocks required to induce latch-up. For inductance values under 17 nH, latch-up is not produced even if all 16 noisy blocks are enabled. As the value of the inductance is increased, the probability of latch-up triggering is higher, and lower activity in the digital section is thus necessary. The conditions for latch-up to occur are thus demanding: the simultaneous switching of a large number of gates is necessary (remembering that every noisy block is made up of 16 buffers), and it must produce noise peaks at the supply lines around 1.5 V, while supplies must be connected to package pins with inductance values over 20 nH. Only some large packages or ones with a high aspect ratio will display these inductances. An accurate pin assignment, choosing the ones in the center to connect the power supplies, should be enough to avoid the problem.

Table 4.II.

Minimum number of noisy switching blocks necessary for noise at the supply lines to trigger latch-up, according to the inductance of the corresponding pins.

Pin inductance (nH)	Minimum number of blocks
40	9
30	11
20	14
17	16

In Figure 4.23 the power supply and NPN transistor base waveforms are shown, with a 17 nH inductance value and 16 and 15 switching blocks. The current through the transistors shows that the SCR structure is excited only in the first case. Similarly, Figure 4.24 shows the same waveforms but now with a 40 nH inductance, and 9 and 8 noisy blocks enabled. Observe that noise levels at the supply lines are similar in both figures, while the main noise difference between 17 nH and 40 nH is the period of the oscillations, which is larger the higher the inductance. A larger period means longer noise pulses, and this facilitates latch-up triggering. The frequency of the oscillations at the power-supply lines may be expressed as

$$f = \frac{1}{2\pi\sqrt{LC}} \qquad \text{Eq. (4.1)}$$

where L is the inductance of the package pins, and C is the equivalent capacitance to ground of the supply line. It can be seen that the ratio of oscillation periods in Figures 4.23 and 4.24 is equal to the square root of the inductance ratio. As for the capacitance C, some simulations were performed modifying the parasitic capacitance values and altering the size of the reference circuit, and it was proved that C depends almost exclusively on the equivalent capacitance to ground of the digital supplies, through the source junctions of the transistors. In other words, the larger the number of transistors powered by a line, the larger the C. This is an important result, insofar as

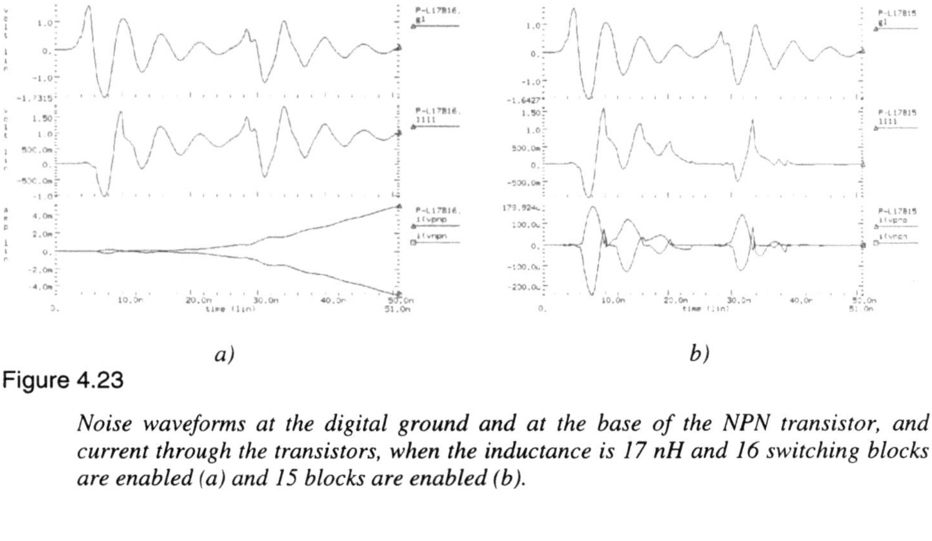

a) *b)*

Figure 4.23

Noise waveforms at the digital ground and at the base of the NPN transistor, and current through the transistors, when the inductance is 17 nH and 16 switching blocks are enabled (a) and 15 blocks are enabled (b).

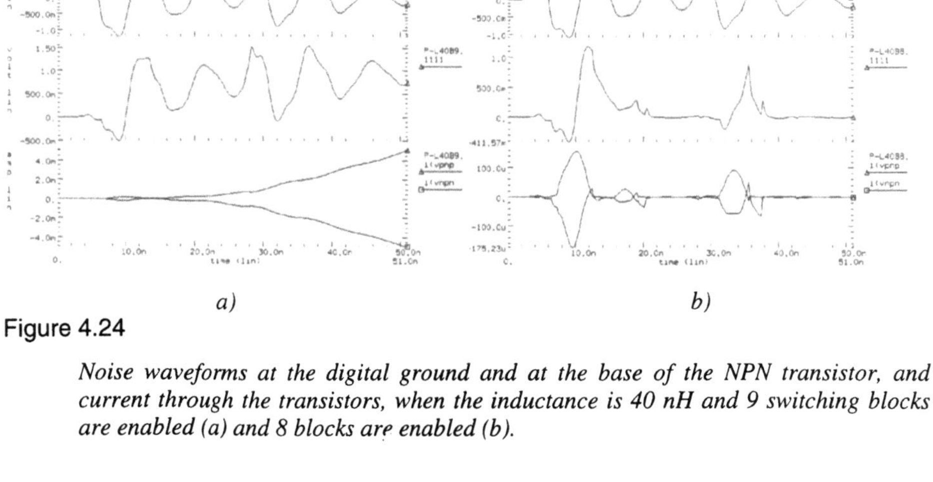

a) *b)*

Figure 4.24

Noise waveforms at the digital ground and at the base of the NPN transistor, and current through the transistors, when the inductance is 40 nH and 9 switching blocks are enabled (a) and 8 blocks are enabled (b).

the latch-up probability for a given level of activity does not depend only on the package inductance, but also on the product inductance per number of gates. In our reference circuit the number of buffers is 256 which, taking into account their size, is equivalent to a circuit containing about 5000 minimum-size transistors. In high complex circuits there are approximately a pair of supply terminals every 50,000 transistors. So in high complexity circuits, the inductance values for which latch-up can be triggered are not as large as our analysis showed, and thus this possibility should be accounted for. It will be useful to bias the substrate at the digital part with the digital supply ground, and thus biasing strategies like E/A, which were advisable from the noise viewpoint, should be discarded in favor of others like D/E or D/A.

4.5 Conclusions

The two main paths through which noise can be coupled to a node in an analog circuit are the lines supplying power to the circuit, and the substrate, basically due to body-effect. Even though we have not studied this, the effect of the first path can be minimized with adequate circuit topologies, like differential ones or ones with high PSRR. Noise propagating in lightly doped substrates is attenuated with distance. Hence, keeping a security distance between noisy circuits and the analog part is in principle enough to minimize the effects of substrate noise. The value of this distance will depend on the amount of noise produced and the analog circuit tolerance to noise. The benefits of distance may be lost if a single line biases all the contacts in the IC, since this way the substrate in both sections is short-circuited. Biasing strategies like A/A or D/D must consequently be discarded. Similarly, and just as important as increasing the distance between analog transistors and digital gates, proximity between contacts connected to the analog supplies and the digital section must be avoided, because noise would get straight into the analog circuitry. Another factor to take into account is the capacitance of the analog supply lines to the substrate, which should be minimized by paying attention to the width of the lines, their routing (avoiding proximity to the digital section), wiring level, or even laying wells under them.

If the former advice is taken, the biasing strategies minimizing noise coupling are D/A, D/E and E/A. The first one minimizes body-effect, although the analog supply can acquire noise through the contacts. E/A strategy also has these characteristics, and moreover noise is reduced because substrate at the digital section is biased with a line not contaminated with switching noise. Nevertheless, this strategy must be discarded because of its high susceptibility to latch-up, particularly in large circuits or ones with much switching noise. The D/E strategy does not involve any of these disadvantages and in general it offers noise levels lower than D/A, although body-effect is more important. As a drawback, it needs an extra supply line, with the consequent increase in area, routing complexity and package pin numbers.

It has been shown that coupling increases with the activity of the noisy part, whereas the effect of the number of substrate contacts depends on the dominant coupling mechanism. If noise is directly coupled through the devices, the more contacts the better, because this way a better return path to ground is provided to the disturbance. On the contrary, if noise is injected by the digital supply lines, or mainly received through the analog ones, it will be advisable to reduce the number of contacts in the respective sections. The ring of pads will not necessarily influence the noise levels at the analog section, whenever analog pads are also placed away from the digital section. Guard rings are effective in this type of wafers, although the noise reduction achieved depends on the characteristics of the layout. Connecting the ring to dedicated biasing lines is recommended, both if D/A or D/E strategies are implemented.

Lastly, it has been proved that the importance of the inductance of the pins assigned to power supplies is greater than the biasing policy. Minimum inductance pins should be chosen, especially for analog and dedicated lines, and if possible multiple pins should be assigned to a single line, thus reducing its effective inductance.

In heavily doped epi wafers, the characteristics of noise propagation and biasing are completely different. The highly conductive part of the substrate allows a very low impedance path between all transistors and contacts, so distance is no longer a factor to take into account. The high number of contacts and large supply line capacitances, especially those of the ring of pads, make impedance between conductive bulk and supply lines very low. If supply lines were connected through an ideal package, biasing would be perfect, all noise would return to ground through the contacts, and the noise coupled would be negligible. Nevertheless this does not occur in practice because of the effect of package parasitics, and the return takes place through both the supplies and other devices. Even backside contacts are not efficient for eliminating noise. Given the usual size of noisy sections, the values of pin inductances required to get good return paths would be very low, under 1 nH, which are difficult to achieve even with multiple assignment. Reducing only the inductance of the backside contacts is also insufficient, in practice. So, from a noise viewpoint, the use of P+ wafers is utterly inadvisable for mixed-signal circuitry implementation.

In this kind of wafers the effect of the pads is larger than in P- substrates, and switching noise coupling from supply lines to the substrate is favored, easily reaching the lines feeding the analog circuitry. The effect of the biasing strategy at the core is therefore much lower because the effect of the periphery is dominant.

References

[1] *Simulation Techniques and Solutions for Mixed-Signal Coupling in Integrated Circuits*, N.K. Verghese, T.J. Schmerbeck, D.J. Allstot, Kluwer Academic Publishers, 1995.

[2] M. Ingels, M.S. Steyaert, "Design Strategies and Decoupling Techniques for Reducing the Effects of Electrical Interference in Mixed-Mode IC's", *IEEE J. Solid-State Circuits*, vol. 32, no. 7, pp. 1136-1141, July 1997.

[3] R. Senthinathan, J.L. Prince, "Simultaneous Switching Ground Noise Calculation for Packaged CMOS Devices", *IEEE J. Solid-State Circuits*, vol. 26, no. 11, pp. 1724-1728, November 1991.

[4] C. Huang, Y. Yang, J.L. Prince, "A Simultaneous Switching Noise Design Algorithm for Leadframe Packages with or without Ground Plane", *IEEE Trans. Components, Packaging and Manufacturing Technology-Part B*, vol. 19, no. 1, pp. 15-22, February 1996.

[5] C.-T. Tsai, W.-Y. Yip, "An Experimental Technique for Full Package Inductance Matrix Characterization", *IEEE Trans. Components, Packaging and Manufacturing Technology-Part B*, vol. 19, no. 2, pp. 338-343, May 1996.

[6] *Influència de les Interconnexions en el Disseny Microelectrònic*, Francesc Moll, PhD Thesis, Department of Electronic Engineering, Universitat Politècnica de Catalunya, Barcelona, 1995.

[7] C.-T. Tsai, W.-Y. Yip, "An Experimental Technique for Full Package Inductance Matrix Characterization", *IEEE Trans. Components, Packaging and Manufacturing Technology-Part B*, vol. 19, no. 2, pp. 338-343, May 1996.

[8] D.K. Su, M.J. Loinaz, S. Masui, B.A. Wooley, "Experimental Results and Modeling Techniques for Substrate Noise in Mixed-Signal Integrated Circuits", *IEEE J. Solid-State Circuits*, vol. 28, no.4, pp. 420-430, April 1993.

[9] J.M. Williamson, M.S. Nakhla, Q. Zhang, P.D. van der Puije, "Ground Noise Minimization in Integrated Circuit Packages through Pin Assignment Optimization", *IEEE Trans. Components, Packaging and Manufacturing - Part B*, vol. 19, no. 2, pp. 361-367, May 1996.

[10] D.B. Estreich, R.W. Dutton, "Modeling Latch-Up in CMOS Integrated Circuits", *IEEE Trans. on CAD of Integrated Circuits and Systems*, vol. 1, no. 4, pp. 157-162, October 1982.

[11] *CMOS Devices and Technology VLSI*, John Y. Chen, Prentice-Hall, 1990.

[12] A. Ochoa, W. Dawes, D. Estreich, "Latch-up Control in CMOS Integrated Circuits", *IEEE Trans. Nuclear Science*, vol. 26, no. 6, pp.5065-5068, December 1979.

[13] M.R. Pinto, R.W. Dutton, "Accurate Trigger Condition Analysis for CMOS Latchup", *IEEE Electron Device Letters*, vol. 6, no. 2, pp. 100-102, February 1985.

[14] S. Odanata, M. Wakabayashi, T. Ohzone, "The Dynamics of Latchup Turn-on Behaviour in Scaled CMOS", *IEEE Trans. Electron Devices*, vol. 32, no. 7, pp. 1334-1340, July 1985.

[15] R.E. Bank, W.M. Coughran, W. Fichtner, E.H. Grosse, D.J. Rose, R.K. Smith, "Transient Simulation of Silicon Devices and Circuits", *IEEE Trans. Electron Devices*, vol. 32, no. 10, pp. 1992-2007, October 1985.

5 Alternative proposals for reducing substrate noise

The focus up to now has been on studying guard rings and substrate bias optimization for reducing noise. We will now analyze the viability of three alternative measures for reducing noise, which may not be advisable for general use due to their cost or drawbacks, but which may be useful in particular cases. First, the use of buried layer in providing a noise return path will be examined. The conditions necessary for maximum efficacy, and the increase in process complexity will be evaluated. Next, we will consider the possibility of producing complementary peaks so they cancel each other out. This technique involves duplicating some circuits – adding dummy circuitry– but even so it would be advisable in the event of noise being produced by a few localized elements. Lastly, we study the viability of using active guards. The technique is similar to the canceling method, but in this case noisy elements are not duplicated, adding instead a circuit which senses noise and creates compensating noise, thus annulling the first one, for instance through negative feedback.

5.1 Buried layers

A highly conductive layer under the switching devices may provide a low-impedance path for noise derivation to ground. Nevertheless, this low-impedance path may also ease coupling between neighboring devices. We will now perform an analysis using the netlist extracted from a simple circuit, similarly to the procedure used in chapter 4, and from the results obtained we will discuss the real advantage offered by buried layers, their optimal distribution and biasing, and their process and technology requirements.

5.1.1. Buried layer design alternatives

We performed the simulation of a circuit consisting of a sensor, and 16 blocks containing 16 noisy buffers each block, synchronously switching. The sensor is a simple amplifier with short-circuited inputs and outputs, loaded with the input capacitances of another identical sensor. A layout of the circuit is generated according to 1.2 μm technology rules, and parasitic capacitances between the nodes of interest can thus be extracted. Figure 5.1 shows a diagram of the circuit. Digital and analog sections use independent supply lines, connected to package pins modeled with the equivalent circuit shown in Figure 4.2.

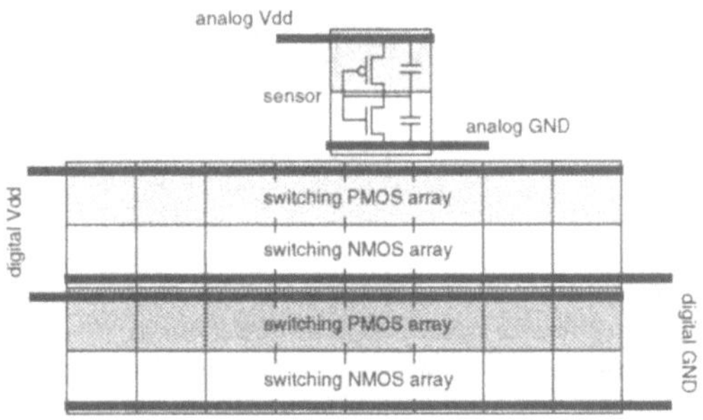

Figure 5.1

Block scheme of the layout generated to extract capacitances. Each of the 16 noisy blocks contains 16 switching buffers.

For the sake of simplicity, the substrate model only considers those resistances between the ports in a noisy block, and coupling resistances between the sensor and the closest noisy block. The values of these resistances are obtained from device simulation of a vertical slice of the coupled blocks –Figure 5.2–, and the extraction is repeated for each of the buried layer configurations considered. It was verified that the only significant resistances are those between every transistor and the biasing contacts, and the resistance between biasing contacts in both sections.

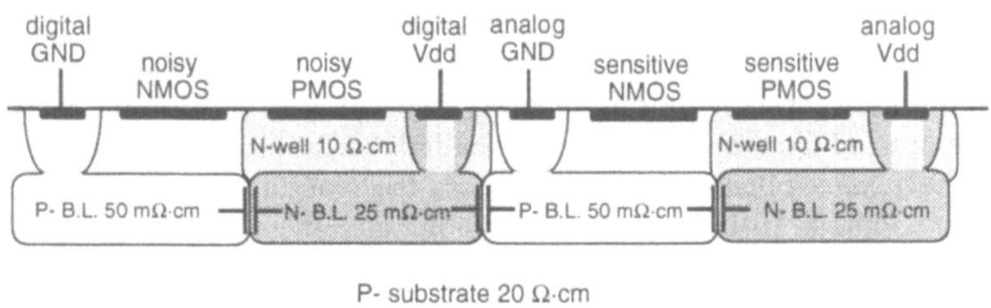

Figure 5.2

Vertical section simulated to obtain the values of the resistance in the substrate model.

First the noise reduction achieved with buried layers is determined, considering three possibilities: 1) placing these just under the noisy section; 2) only under the sensitive devices; or 3) under all the circuitry. Moreover, the need to bias the buried layers through deep contacts is also evaluated. In the areas with no buried layers, substrate is biased with conventional contacts. In all cases, bias contacts are connected to supply voltages also powering the circuitry in the corresponding section. The results obtained are shown in Table 5.I.

Table 5.I.

Comparison of the noise levels obtained in the different possible implementations of buried layers.

	with deep bias contacts	without deep bias contacts
without buried layers	—	248 mV_{pp}
buried layers only under the digital section	105 mV_{pp}	183 mV_{pp}
buried layers only under the analog section	75 mV_{pp}	232 mV_{pp}
buried layers under all the circuitry	26 mV_{pp}	165 mV_{pp}

If the results are correlated with the resistances in the substrate model, the importance of the layers can be appreciated. The presence of a buried layer under analog circuitry reduces the resistance between the sensor and the ground line by an order of magnitude, in this particular example. This produces a good biasing of the substrate under the sensor –contacts are connected to a line free of noise–, and then we get a minimum amount of noise. The buried layer under the digital part the coupling resistance between the sensor and noisy gates increases by a factor of four, while also reducing the amount of noise in the substrate under the digital section. The noise level arriving at the analog section is consequently reduced. The combined effect of both buried layers drastically increases the insulation between the digital and the analog part, achieving in our example a 90% noise reduction. This result is obtained in spite of the capacitance appearing between the N-type digital layer and the P-type analog layer, and despite the low resistance between the bias contacts in both sections, which increase the noise level at the analog ground line.

It can also be observed that the effect of the deep bias contacts is essential, because otherwise the layers become a path that allows easy noise transmission, similarly to what occurs in a P+ substrate without grounded backplane.

Figure 5.3 compares the noise waveforms at the sensor according to the presence of the buried layer under the digital circuitry. These waveforms are independent of the noise level, and their difference is due to the resistive part of the impedance between digital supply lines. When extracting the capacitances in a circuit, there is a large capacitance between *GND* and V_{dd} digital supplies –the same in analog ones–, basically due to the supply lines parasitic capacitances to wells and substrate, and to the junctions of the wells. In order to adapt to the real situation, a series resistance must be connected to this capacitance, standing for the resistance of the substrate to the biasing contacts. In the event of a buried layer existing, this resistance is orders of magnitude smaller. When a current peak is produced, the resonant circuit formed by the package inductances and the capacitance between supply lines produces oscillations in these lines, which are transmitted to any node in the circuit powered by these lines. In lightly doped substrates, where the series resistance has a large value, the oscillations are rapidly damped, but this is not the case when buried layers are used. The value of the time constant will depend on each particular circuit, determined by size and technological characteristics, but given a circuit, the oscillations damp more slowly –time constants 2 or 3 orders of magnitude larger–if there are buried layers under the digital circuitry.

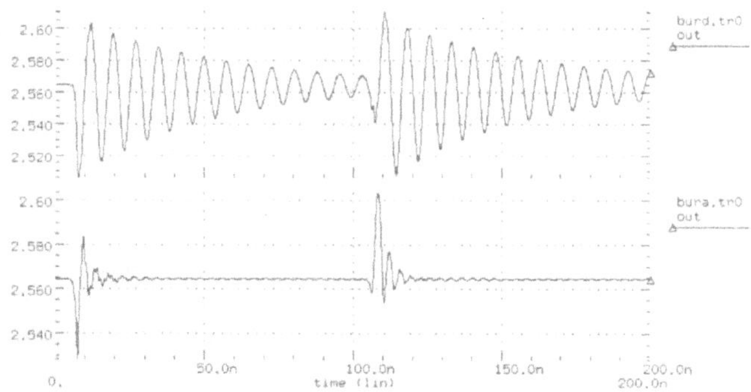

Figure 5.3

Noise waveforms at the sensor when buried layers are placed under the digital section (up), and when no buried layers are present (down).

Given that the best alternative is to implement buried layers both under the digital and the analog section, simulations were also performed to determine where to connect those buried layers. It was seen that using the same line to bias all buried layers is a bad option, even if this line is exclusively dedicated. The best options are those also pointed out in the previous chapter: the layer under the analog section can be biased with the analog ground, while the layer under the digital section is better biased with a low-impedance dedicated ground.

Lastly, Figure 5.4 shows the noise dependence on the package inductance of the model shown in Figure 4.2. It can be seen that noise is significantly reduced with the inductance also when using buried layers, although for a given inductance value, the noise reduction achieved in respect of a conventional P- substrate continues to be around 90%.

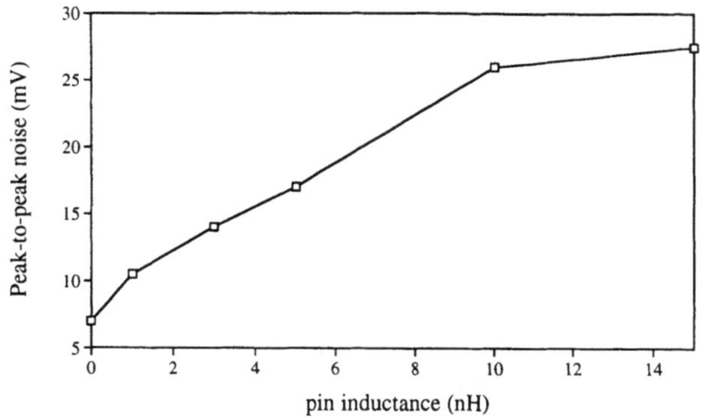

Figure 5.4

Noise levels at the sensor according to the pin inductance, when buried layers are placed under both sections.

5.1.2. Technological aspects

The resistive model used in the simulations showed that the advantage of buried layers lies in the reduction of the resistance between transistors and contacts, and through increasing the coupling impedance between transistors. Nevertheless this is not accomplished if a low resistive path is allowed between the buried layers under the digital and the analog sections. Thus, in order to guarantee effectiveness, it is important to avoid ohmic junctions between them, unless these are produced through the highly resistive substrate. Remember that buried layers are also an efficient means of reducing latch-up, so it is pointless to implement them on heavily-doped wafers.

Figure 5.5 shows the pattern for placement in a mixed-signal circuit to guarantee that buried layers under different sections have capacitive junctions, never low resistive ohmic junctions. The separation between sections may also be increased for better insulation. Nevertheless, a P+ buried layer –on a P wafer– cannot be placed in this region. Then, the insulation area should therefore be highly resistive and contain no buried layers –meaning that buried layers could not be implemented in a self-aligned process–, or have an N+ layer on a P wafer –which would create PN junctions, thus increasing insulation–.

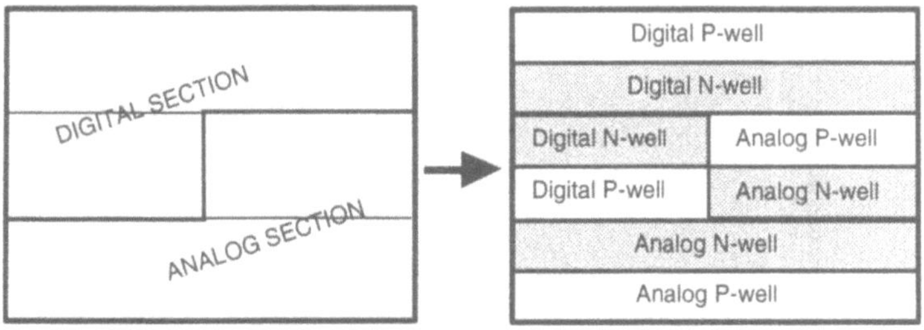

Figure 5.5

Recommended placement in a mixed-signal circuit like the one on the left, to achieve efficient buried layers for reducing substrate coupling.

The implementation of buried layers entails significantly greater complexity in the manufacturing process. At present buried layers are commonly used in high performance BiCMOS processes. These processes typically contain vertical NPN bipolar transistors, and N+ buried layers under the N-well acting as a collector must be implemented in order to get reasonable collector resistance values. Furthermore, a deep N+ contact is usually included to further reduce the collector resistance. In order to save masks, buried layers are also placed under the N-wells in the PMOS arrays, with the added advantage of lower susceptibility to latch-up, and meaning no need for P+ wafers. The light doping of P- wafers is insufficient to get proper insulation between bipolar transistors and avoid punch-through, so a large separation is needed between bipolar devices. In order to increase integration density, advanced processes also include self-aligned P+ buried layers, which are used as insulation and again are also implemented under the NMOS transistors. These P+ buried layers do not need any biasing or special contact. Figure 5.6 depicts a vertical section where the device structure in a high performance BiCMOS process can be seen.

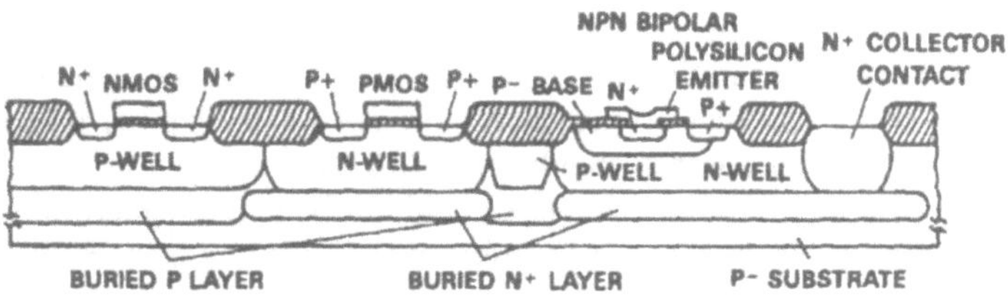

Figure 5.6

Device structure in a high performance BiCMOS process. It includes self-aligned N and P buried layers, N and P wells independently adjusted, deep collector contact and polysilicon emitter [1].

CMOS process	BiCMOS process
* *self-aligned buried layers*	* self-aligned buried layers
epi layer deposition	epi layer deposition
* N-well	* N-well / collector
* P-well	* P-well
* active area definition	* active area definition
* isolation	* isolation
field oxide	field oxide
* *deep N+ contacts*	* deep N+ contacts
* *deep P+ contacts*	* *deep P+ contacts*
	* base implant
gate oxide	gate oxide
adjust implant	adjust implant
* poly	* poly
* LDD	* LDD
spacer	spacer
* N+ S/D implant	* N+ S/D/emitter implant
* P+ S/D implant	* P+ S/D/base implant
* contact	* contact
* metal 1	* metal 1
* via	* via
* metal 2	* metal 2
passivation	passivation
* pad windows	* pad windows

Figure 5.7

Summary of the manufacturing steps in typical CMOS and BiCMOS processes [1], [3]. The asterisk (*) marks the steps requiring new masks. The steps that should be added to implement buried layers to reduce substrate coupling are given in italics.

Starting from a BiCMOS process with the characteristics described, the inclusion of buried layers to reduce substrate coupling would not imply a large increase in process complexity. The only extra step is to add the implant to make the P+ deep contacts, which were important enough according to the results in Table 5.I. This means adding a new mask to the 16 ones usually used by these processes. On the other hand, in a conventional twin-well CMOS process, the number of steps to add would be much larger: implementation of the buried layers, epitaxial growth, and the N+ and P+ deep contacts. For a number of masks, this means adding 3 extra ones to the 13 masks typically used by these processes. The extra cost increment that this involves is significant, taking into account that some of the new process steps may be considered critical ones. Taking into account the yield reduction produced by the faults associated with these new steps, the implementation of buried layers would mean a cost increment between 25% and 35% for a CMOS process [1]. This cost increase may furthermore not be the most important drawback, compared with the loss of integration density. Lateral diffusion of the deep implants, which must be implemented in the first stages of the process, would increase the minimum distance between contacts and devices, making a general increase in the area of a circuit. In spite of all these disadvantages, buried layers have already been successfully used in some mixed-signal circuits for increasing the insulation between sections [2].

The effectiveness of buried layers will also depend on the technological characteristics of their implementation, particularly those connected with the resistance they offer to the biasing contacts –layer resistivity, depth, and thickness–. Figure 5.8 shows the noise dependency on the resistivity of the buried layer, for the particular case analyzed, and it can be seen that a value between 2 and 3 orders of magnitude smaller than that of the substrate is suitable to get good results.

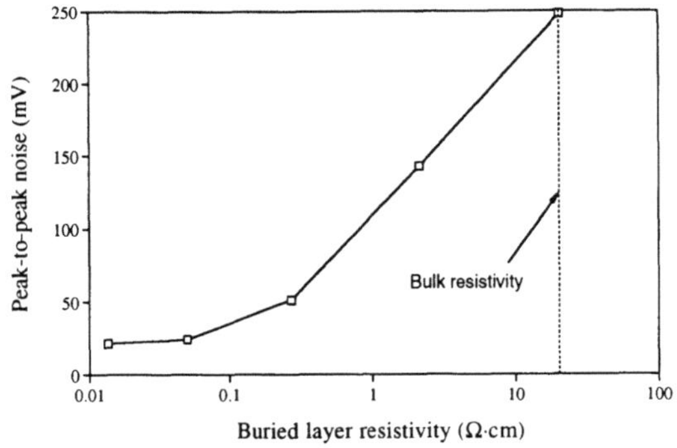

Figure 5.8

Noise level according to the resistivity of the buried layers.

5.2 Noise compensation

In some circuits, substrate noise is originated by certain particular items like large drivers, especially those of the digital output pads, which usually toggle simultaneously. If we accept that the noise produced by a large transistor is mainly injected to the substrate through the junction capacitances (see chapter 2), it will be true that the noise peak produced by a 1 to 0 logic transition has a sign opposite to the noise peak produced by a 0 to 1 transition. This is the basis for the idea of noise compensation: adding dummy devices, counter-phase switching in respect of the original noisy ones, so that the noise produced by both switchings is canceled out. In this section we will determine whether noise compensation is really feasible, and what the conditions required to get a really efficient dummy element are.

The analysis is performed on a circuit like the one considered in the former section, with a noisy section consisting of 16 blocks containing 16 noisy elements each, as shown in Figure 5.1. The difference is that now half of the blocks are switching counter-phase in respect of the other half. Switching is synchronous, so the circuit can be representative of large drivers. The parasitic capacitances and pin models are the same as those in the former section, while the resistance extraction is performed assuming the characteristics of P- or P+ substrates. For the sake of simplicity, only the coupling net between the sensor and two switching blocks is considered, one of them switching counter-phase. The aim of the rest of the blocks is to introduce switching noise at the supply lines. Before analyzing the compensation, some previous simulations are made to choose the dimensions of the disturbing elements, so the value of the noise peak produced by both transitions is the same.

The results of the analysis on lightly doped P- substrates are first presented. Figure 5.9 shows the total noise results for three different package inductance values. The horizontal axis contains the time skew between the switching of both blocks. The noise levels present large differences depending on the inductance value, so results are presented related to the noise produced by 8 switching blocks, with no compensation.

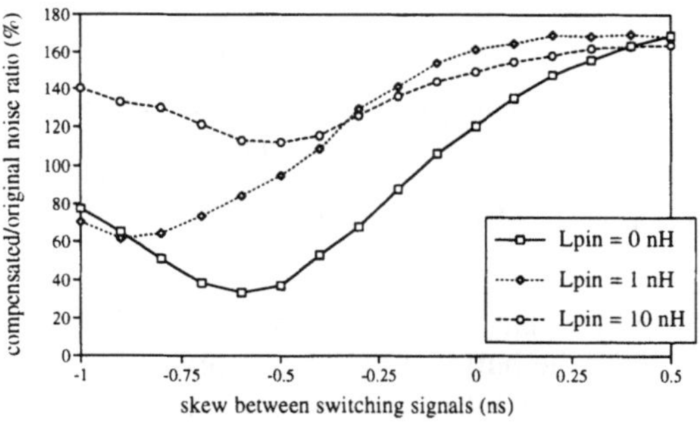

Figure 5.9

Noise canceling according to the skew between counter-phase switching buffers, P-substrate.

When analyzing the results in Figure 5.9, it is first observed that the maximum noise reduction achievable depends largely on the value of the pin inductances. Although in ideal conditions we could reduce noise to 30% of the original level, with an inductance of only 1 nH, the noise level is already 60% of the original one, and with a 10 nH no noise reduction is achieved, but an increase, in fact. The reason for this behavior can be understood by observing in Figure 5.10 the waveforms individually produced by the noisy and compensating blocks, with inductance values 0 and 10 nH. We can see that in the ideal case, with no noise at the supply lines, the waveforms resulting from both transitions are complementary, so the total effect is noise cancellation. Nevertheless, if the inductance has a significant value, noise at the supply lines is important enough, so the disturbance is mainly introduced to the substrate through the contacts, and the waveforms resulting from each transition are not complementary, but practically identical, thus making noise cancellation unfeasible.

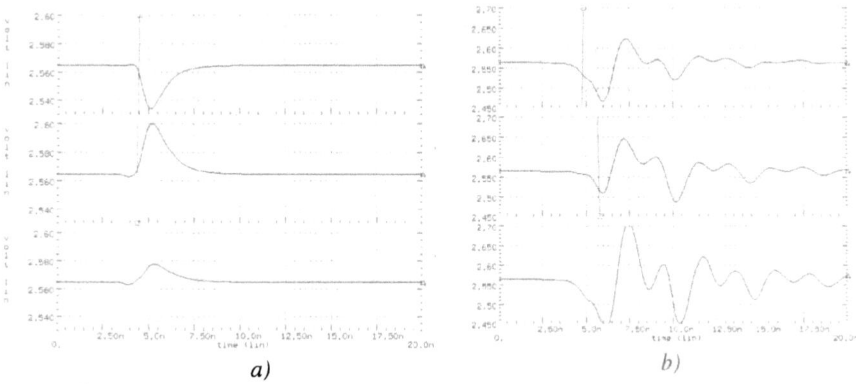

Figure 5.10

Noise waveforms individually produced by noisy and compensating blocks (up and middle), in complementary transitions, and noise waveform produced by the simultaneous switching of all blocks (down). a) When supply lines are free of noise. b) With package pins modeled with 10 nH inductances.

We consequently conclude that noise compensation will only be feasible when the disturbance introduced by the buffers through the device junctions is much larger than the noise in the supply lines introduced through the substrate contacts. This will depend both on the package characteristics and on the activity of the digital section and layout characteristics, so it will be difficult to know a priori if noise cancellation will be feasible, although we must admit that most of the situations will be unfavorable. This is even truer in P+ substrates where, as can be seen in chapter 4, substrate noise is mainly introduced by power-supply lines, making compensation even less feasible. The results obtained in P+ substrates are shown in Figure 5.11.

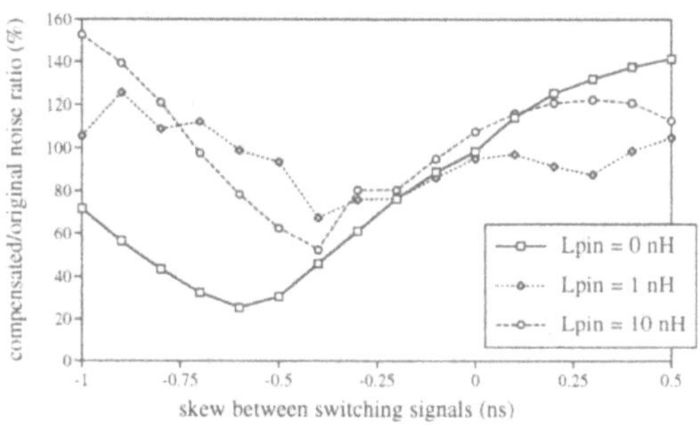

Figure 5.11

Noise canceling according to of the skew between counter-phase switching buffers, P+ substrate.

Nevertheless this is not the only difficulty that comes up when trying to get noise cancellation. Going back to Figures 5.9 and 5.11, we can see that the synchronization requirements between both transitions are really demanding, and a skew of a few hundred picoseconds may lead to a noise increase instead of a cancellation. This is because noise has frequency components in the range of hundreds of megahertz and gigahertz which produce short noise pulses. The practical consequence of these synchronism demands is the low feasibility of noise compensation. From a given signal it will be enormously difficulty to generate a counter-phase signal with transients perfectly synchronous to the first one, and with a very small tolerable skew. The delay conditions not only depend on the signal driving the buffers, but also on the loading conditions. Thus, it is not enough to design compensating elements with identical characteristics to those of the original ones, but their fan-out must also be the same. This is especially difficult to achieve in output buffers, which were the items seen to be most suitable for applying the noise compensation technique.

In connection with this aspect, it will also be necessary for both transitions to produce the same noise level, or at least very similar ones. This requirement will determine the characteristics of the buffers, which should be designed to get the same rise and fall times.

The conclusion reached from the analysis put forward above is that the implementation of compensating elements to produce significant noise reduction involves some practical difficulties that make noise compensation an unattractive and inefficient technique. Other techniques presented in the literature for reducing the amount of noise produced by output buffers, like the limitation of the slew-rate [4], probably provide the same or better results more soundly, avoiding the dependencies on the physical and electrical characteristics of each particular circuit. Furthermore, the effect of noise in the supply lines and the oscillations produced by the parasitic capacitances and inductances make the maximum possible noise reduction really low. In spite of these drawbacks, noise compensation has been proposed in [5], in this case to cancel the ringing produced by output buffer switching, and thus increase the noise margin of these signals. The waveform of the ringing is a damped oscillation, and the authors' proposal is to produce two delayed switchings so that the maximum peak produced by one of the switchings is coincident with the minimum produced by the other one. It must be observed that this means there are not two complementary waveforms, so a 100% cancellation could never be achieved. The authors meet with some other difficulties that we have also found, like the small tolerance to switching skew, and the difficulty in fixing this skew, which depends on the period of the oscillations which in turn depends on the multiple parasitic elements in the circuit, package and load. The authors solve this problem by making the time skew adjustable with an external voltage, which should be adjusted to get maximum cancellation. This solution is hardly feasible, since every output signal would then require its adjustment voltage. The experimental results presented by the authors show a poor noise margin improvement.

5.3 Active guard bands

In the noise compensation technique presented in the previous section, it is necessary to know in advance which elements are the noisiest ones, and add as many dummy elements as noisy ones, which may easily lead to duplicating a large part of the circuit. An intuitive evolution of the noise cancellation concept to all circuits leads one to think on a circuit that senses noise waveforms and originates a complementary waveform so both cancel each other out. This idea has been proposed in [6] in a paper including experimental measurements. The authors propose the implementation of the canceling circuit or active guard with an operation amplifier (OA) with negative feedback, particularly an inverting amplifier configuration. The inverting input and the output of the OA are connected to the substrate with ohmic contacts, so the resistive substrate acts as a feedback loop and as an input resistance. This can easily be seen in a P+ substrate, as shown in Figure 5.12.

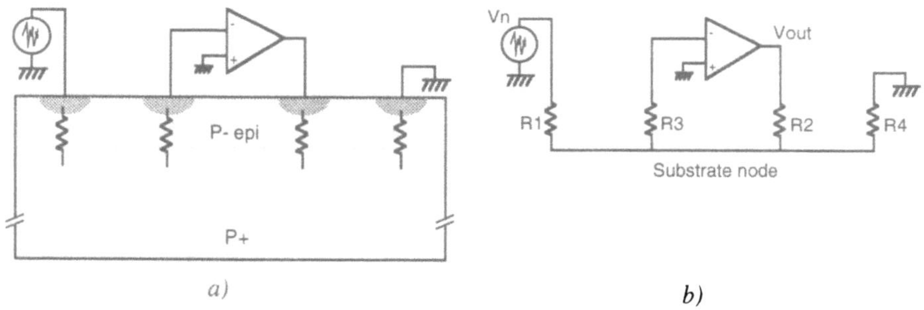

Figure 5.12

Diagram scheme of the connection of an active guard (a) and equivalent circuit (b) [6].

The scheme of the active guard, substrate contacts, and noise injection, is shown in Figure 5.12.a, assuming that the noise is introduced at a given point, and skipping the effect of the rest of the circuitry in the chip. If we assume that the P+ bulk behaves like a single node, and the substrate is modeled only with the resistances through the epitaxial layer, which is a reasonable assumption, the equivalent circuit in Figure 5.12.b is obtained. This model is formed by the resistances R_1, between the noise injecting node and the substrate, R_2 for the contact connected to the output of the OA, R_3 for the contact connected to the OA inverting input and R_4 as the equivalent resistance to the substrate contacts.

Assuming zero current at the OA inputs, and an open-loop gain A, the voltage at the output of the amplifier is

$$V_{out} = -V_n \frac{A}{1+\frac{R_1}{R_2}(A+1)+\frac{R_1}{R_4}} \qquad \text{Eq. (5.1)}$$

while the substrate noise is given by

$$V_{substrate} = V_n \frac{1}{1+\frac{R_1}{R_2}(A+1)+\frac{R_1}{R_4}} \qquad \text{Eq. (5.2)}$$

where V_n is the noise at the source. If the gain of the OA is high enough, equation (5.1) can be approximated by

$$V_{out} = -V_n \frac{R_2}{R_1} \qquad \text{Eq. (5.3)}$$

which is the elementary expression of an inverting amplifier. By applying superposition, the former expressions may be generalized for a large number of noise sources. From (5.3) we can deduce that, if a compensating signal symmetric to the noise introduced is desired, the condition $R_1=R_2$ must be accomplished, i.e., the resistance between the contact of the active guard and the bulk must be equal to the equivalent resistance between noise injectors and substrate. This is in practice hardly feasible because of the difficulty involved in predicting the values of these resistances, which will be different for every noise source. Nevertheless, in the circuit proposed the noise reduction is not achieved by cancellation, as stated in [6], but because the negative feedback forces a zero voltage at the substrate node, since in fact the voltage at this node equals the voltage at the negative input of the OA. Consequently, a resistance ratio that makes the amplifier work in its non-saturated region is enough to get noise reduction.

The authors of the work show an experimental approximation to their technique with a test circuit consisting of a number of contacts connected to an off-chip operational amplifier. They use a sinusoidal wave as a noise signal, and they get noise elimination for frequencies up to 1 MHz, and a lower but still significant reduction up to 20 MHz.

Although the theoretical setting of the technique is attractive and has great potential, its efficacy may in practice be limited by technological setbacks or non-idealities. In this section we will make an analysis of the limitations that would appear when applying the technique to a realistic situation, the real possibilities of the technique, and the different implementation alternatives.

The first limitation to take into account is the bandwidth. In previous chapters it has been seen how substrate noise waveforms display fast peaks, which lead one to think of a noise spectrum centered at high frequencies. If this is true, high performance amplifiers will be needed, and even these may have limited effectiveness, in the best of cases making the circuit more expensive. To get an idea of the frequency components of substrate noise, we obtain the spectrum of the noise waveforms when simulating a circuit like the one presented in the previous sections, composed of a number of noisy blocks, and considering the different parasitic elements associated with supply lines and package. In the analysis presented in this chapter and in the previous one we have always obtained waveforms with a similar appearance, so we take them as a representative example. The spectrum is obtained by applying a fast Fourier transform (FFT) to the noise waveform, with a rectangular window. Figure 5.13 depicts the waveforms in the time domain and their spectrums, both for the noise at the substrate node in a P+ wafer, and for the noise received by an analog amplifier acting as a sensor. It can be verified that the frequency content is important up to 3 GHz, with a main peak at 800 MHz, and a secondary peak at 2.1 GHz. In any event, the largest components lie around a few hundreds MHz. The frequency content, particularly the position of the two most important peaks, may vary according to the parasitic elements in each particular circuit, but it will be located in the same range.

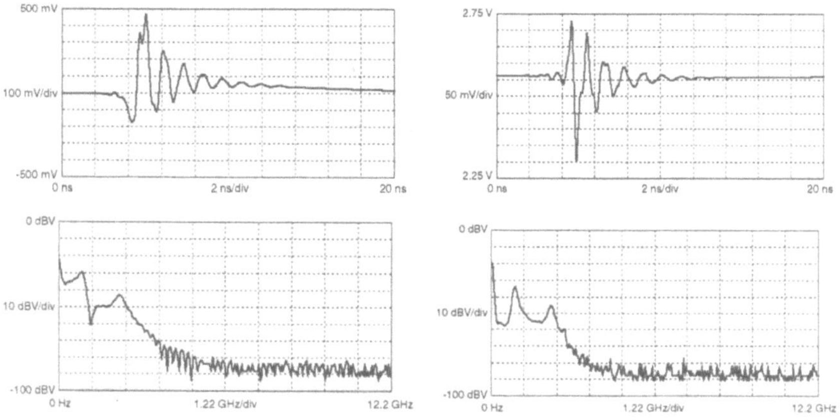

Figure 5.13

Time and frequency noise response at the substrate node in a mixed-signal IC (a), and at the output of an analog amplifier (b).

From the spectral analysis it can be determined that, if an active guard band technique is to be used for noise reduction, operational amplifiers with a large bandwidth are needed. Most commercial operational amplifiers have low bandwidth, and even very high frequency amplifiers would be insufficient to filter a large part of the noise spectrum. Figure 5.14 shows the frequency response at the substrate node in the former circuit, when the active guard is implemented with some commercial OA with different nominal bandwidths. The digital ground line has been used as the noise-injector node, and the parasitics of the package pins connecting the substrate contacts to the off-chip amplifier have been considered, too. It can be seen that even OAs with gain-bandwidth products larger than 500 MHz do not provide noise elimination over 100 MHz. For frequencies up to 10 MHz, the results offered by active guards are excellent, so we could consider using this technique in the case of substrate noise not being produced by digital switching gates, but by some medium-frequency analog circuit. In any event, this situation has not been described in the literature available.

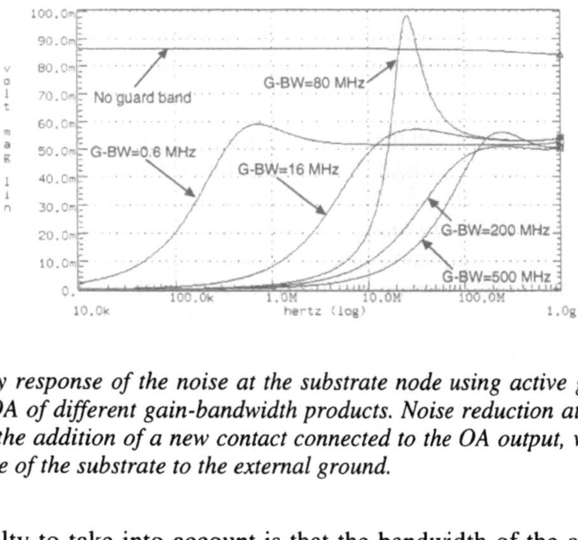

Figure 5.14

Frequency response of the noise at the substrate node using active guard bands with off-chip OA of different gain-bandwidth products. Noise reduction at high frequencies is due to the addition of a new contact connected to the OA output, which reduces the impedance of the substrate to the external ground.

An extra difficulty to take into account is that the bandwidth of the active guard will be affected by its gain as an inverting amplifier. A unity gain is desired in order to maximize the bandwidth, i.e. equal resistances R_1 and R_2 in Figure 5.12. This is difficult to achieve, especially because of the difficulty in getting a good estimation of R_1, unless the noise is introduced by a single element. Once an estimation of R_1 was obtained, a contact giving a suitable value of R_2 would be needed. Moreover, typically $R_1 << R_2$, which may drive the amplifier out of its linear region. Figure 5.15 shows the noise frequency response at the substrate node, when an active guard is included, with different values of R_2 that provide different gain values. It can be seen that gains larger than unity give decreasing bandwidths, while a value of R_2 some tens larger than R_1 drives the guard out of operation.

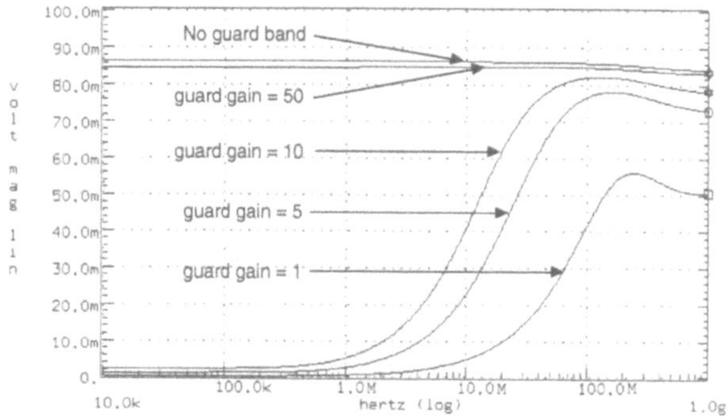

Figure 5.15.

Frequency response of the noise at the substrate node when an off-chip active guard is used, according to its gain.

The inefficiency of active guards in eliminating the substrate noise originated by digital switching is lastly proved by a time-domain simulation of a mixed-signal circuit on a P+ wafer. Figure 5.16 depicts the waveforms at the substrate node with and without an active guard. In the upper curves the noise signal at the ground line is a sine wave of 1 MHz frequency, and the active guard eliminates noise almost completely. In the middle curves the noise is a 50 MHz sine wave, and the noise reduction achieved is already rather insignificant. Moreover, it can be seen that the response time of the OA is in the order of the noise period. In the lower curves, noise is produced by digital switching gates, and the active guard has hardly any effect.

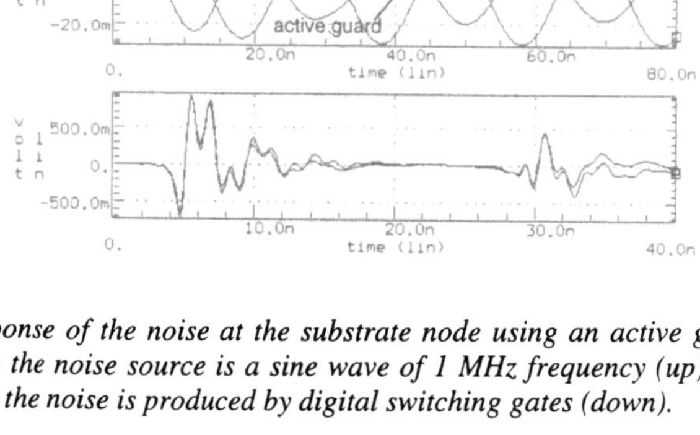

Figure 5.16

Time response of the noise at the substrate node using an active guard with off-chip OA, when the noise source is a sine wave of 1 MHz frequency (up), 50 MHz (middle) and when the noise is produced by digital switching gates (down).

Both in the work presented in [6] and in the results presented so far in this section, off-chip operational amplifiers have been assumed. If the limitations that prevent the practical usage of this technique were in some case overcome, it would be attractive to integrate the operational amplifier in the same noisy chip. Nevertheless the implementation of on-chip active guards involves some further difficulties. The origin of these drawbacks is that the negative power-supply of the OA is the voltage that biases the substrate, *GND*. If the circuit is to work in its linear region, DC values at the inputs of the amplifier should be placed somewhere between the supply voltages. If the circuit is powered at 0 - 5 volts, like in most mixed-signal technologies, input voltages would be typically 2.5 V. This means that a capacitor to couple substrate noise to the OA input will be needed, and another capacitor will similarly be needed at the OA output. The equivalent circuit in this situation is thus the one shown in Figure 5.17.

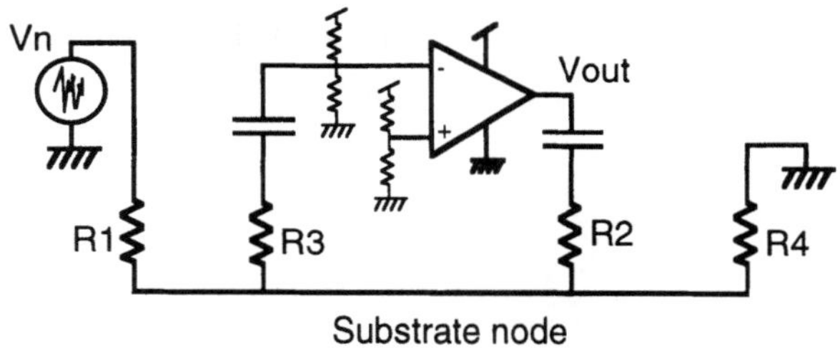

Figure 5.17

Equivalent circuit for an on-chip active guard.

The capacitance at the feedback loop now makes a voltage at the output of the amplifier equal to

$$V_{out} = -V_n + \frac{A}{1+\frac{R_1Cs}{1+R_2Cs}(A+1)+\frac{R_1}{R_4}} \qquad \textit{Eq. (5.4)}$$

which, if the OA gain is high enough, can be approximated by

$$V_{out} = -V_n\frac{1+R_2Cs}{R_1Cs} \qquad \textit{Eq. (5.5)}$$

in other words, the active guard band behaves like an inverting amplifier just when $R_2Cs>>1$, otherwise the gain is high and the amplifier is quickly driven to saturation. This implies that capacitor values should match the frequencies of interest, both to fulfill the former condition and to offer a low enough impedance.

Taking the reference circuit previously used, and using a standard cell OA from the libraries of the corresponding technology –1.2 µm–, time-domain simulations are made to determine the conditions necessary to achieve maximum noise reduction at different frequencies. With a noise waveform of 10 KHz frequency, the on-chip active guard can reduce substrate noise to 7% of the original value. With a 100 KHz frequency noise is reduced to 12%, but the capacitances necessary to get these noise levels are 500 nF and 50 nF, respectively. These values make use of on-chip active guards unfeasible at these frequencies. Smaller capacitors at the OA input and output contacts allow the bias current and the loop gain to drive the amplifier out of its linear region. At higher frequencies the capacitances take on values that are prone to integration, but then we face the problem of the amplifier bandwidth. So, at 1 MHz, the noise is just reduced to 65%, while at 10 MHz 89% of the original noise still remains. The amplifier used has a unity-gain bandwidth of 11 MHz, and although higher bandwidths may be achieved in more advanced technologies, they are still insufficient to eliminate switching noise. Figure 5.18 shows the effect of an on-chip active guard on the noise at the substrate node at different frequencies, and when caused by switching noise.

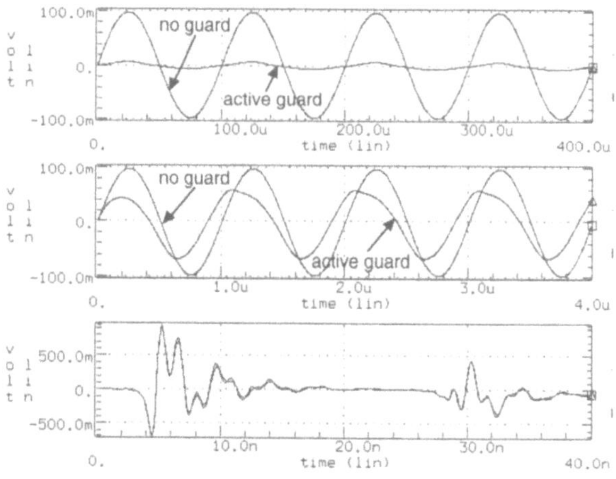

Figure 5.18

Time response of the noise at the substrate node when using an on-chip active guard, when the noise source is a sine wave of 10 KHz frequency (up), 1 MHz (middle), and when the noise is produced by digital switching gates (down).

References

[1] *BiCMOS Technology and Applications*, 2nd edition, A.R. Alvarez editor, Kluwer Academic Publishers, 1993.

[2] M. Zuffada, R. Alini, M. Demicheli, M. Gregori, D. Moloney, S. Portulari, F. Sacchi, S.O. Arf, V. Condito, R. Castello, "A Single-Chip 9-32 Mb/s Read/Write Channel for Disk-Drive Applications", *IEEE J. Solid-State Circuits*, vol. 30, no. 6, pp. 650-659, June 1995.

[3] *Atlas of IC Technologies: An Introduction to VLSI Processes*, W. Maly, The Benjamin/Cummings Publishing Company, 1987.

[4] R. Senthinathan, J.L. Prince, "Application Specific CMOS Output Driver Circuit Design Techniques to Reduce Simultaneous Switching Noise", *IEEE J. Solid-State Circuits*, vol. 28, no. 12, pp. 1383-1388, December 1993.

[5] T. Sekiguchi, M. Horiguchi, T. Sakata, Y. Nakagome, S. Ueda, M. Aoki, "Low-Noise, High-Speed Data Transmission Using a Ringing-Canceling Output Buffer", *IEEE J. Solid-State Circuits*, vol. 30, no. 12, pp. 1569-11574, December 1995.

[6] K. Makie-Fukuda, S. Maeda, T. Tsukada, T. Matsuura, "Substrate Noise Reduction Using Active Guard Band Filters in Mixed-Signal Integrated Circuits", *Symposium on VLSI Circuits Digest of Technical Papers*, pp 33-34, 1995.

6 Experimental measurements performed on a mixed-signal IC

This chapter deals with experimental measurements performed on a mixed-signal test IC. These measurements allow comparison of noise performance of lightly and heavily doped wafers. The results obtained will also be used to verify the results obtained in previous chapters, particularly the ones to do with dominant coupling paths, and to validate the models used in chapter 4. Our aim is also to prove how prone circuits like A/D converters and RAM cells are to substrate noise. Some comparators commonly used by these circuits have been included in the IC (auto-zeroed, sense amplifier), and the drop in resolution produced by substrate noise will be measured.

6.1 Description of the test circuit

6.1.1. General characteristics

The test IC designed is manufactured with a CMOS 1.2 μm twin-well mixed-signal technology, P substrate, two metal layers and two polysilicon layers that will facilitate capacitor implementation. The technology is intended for circuits to be powered at 5 V. The most relevant parameters of the transistors are summarized in Table 6.I.

The same circuit is implemented on two kind of wafers: P- with a typical uniform resistivity of 8.5 Ω·cm, and P+ with a 0.7 mΩ·cm resistivity and an epitaxial layer about 12 μm thick and with resistivity 8.5 Ω·cm. The wafer thickness is 525 μm, and a metal layer is fitted on the backside to act as a substrate bias contact. Chip dimensions are 1700 × 2750 μm^2, and its layout is custom designed except for the pads. The scribe lane usually implemented at the edge of any IC has been missed out in order to avoid its influence on the biasing. The chip periphery is made up of 18 digital input pads (corresponding to control signals and the switching clock); 21 analog pads with a 100 Ω series resistance (connected to the outputs and control signals of the sensors); and 9 supply pads distributed the following way: two pads to power the digital

Table 6.I.

Parameters of the transistors in the process used.

Parameter:	NMOS	PMOS	units
minimum gate length (drawn)	1.2	1.2	μm
gate oxide thickness	245	245	Å
threshold voltage	0.67	-0.7	V
gain	68	26	$\mu A/V^2$
body effect factor	0.66	0.5	$V^{1/2}$
substrate effective doping	$2.6 \cdot 10^{16}$	$1.5 \cdot 10^{16}$	cm^{-3}

periphery, two for the digital core, another two for the analog circuitry, and two ground and one V_{dd} pads exclusively dedicated for biasing purposes. There is a total of 48 pads, so the circuit is packaged in a DIL-48S. A conductive epoxy is used to glue the chip to the package metal backplane, which is wire-bonded to the two dedicated ground pins. These pins thus have double wiring, one to the pad and another to the base of the chip. The criterion followed in pin assignment attempted to minimize the inductance of power-supply lines.

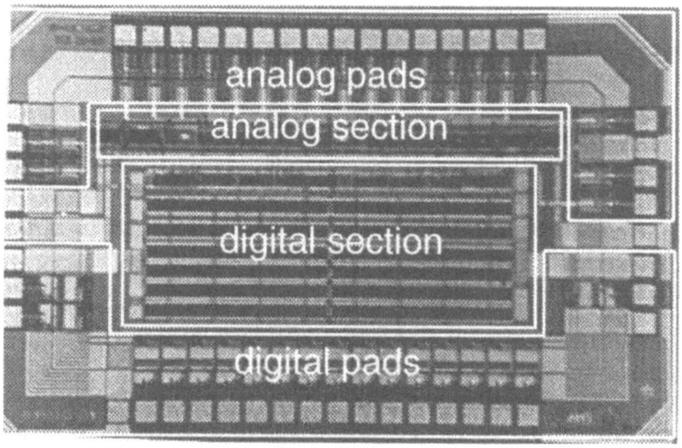

Figure 6.1

Photograph of the chip designed to obtain noise measurements.

Given that noise produced depending on biasing strategy is to be measured, the power lines feeding the transistors are different to the ones biasing the wells and the substrate. In each section one of the three available options can be chosen: analog supply, digital supply, or a dedicated line. Supply line selection is done by means of large switches, with ON resistances under 50 Ω.

The analog section is enclosed by a P+ guard ring which can also be connected to any of the three ground lines, or leave the connection in high impedance. In order to minimize the series resistance, selection of the ring biasing is performed by two identical switches, which is allowed by the symmetry of the layout.

6.1.2. Description of the digital section

The digital section consists of a noisy section with controllable activity, and synchronization circuitry. The noisy section is structured in the form of a number of identical blocks which may be enabled by control signals. Every block contains 16 simultaneously switching buffers, with 0.5 ns rise time and 0.4 ns fall time. The schematic of one of the blocks is shown in Figure 6.2, including transistor dimensions, while Figure 6.3 shows its layout. The clock signal reaching these blocks has gone through an adjustable delay circuit in order to synchronize buffer switching with the sampling of the analog comparators.

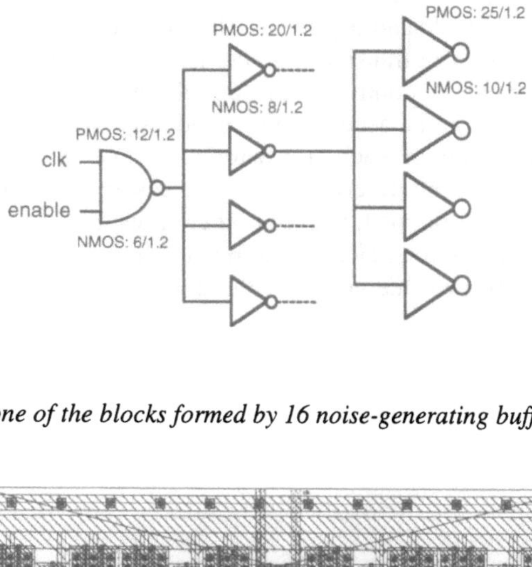

Figure 6.2

Schematic of one of the blocks formed by 16 noise-generating buffers.

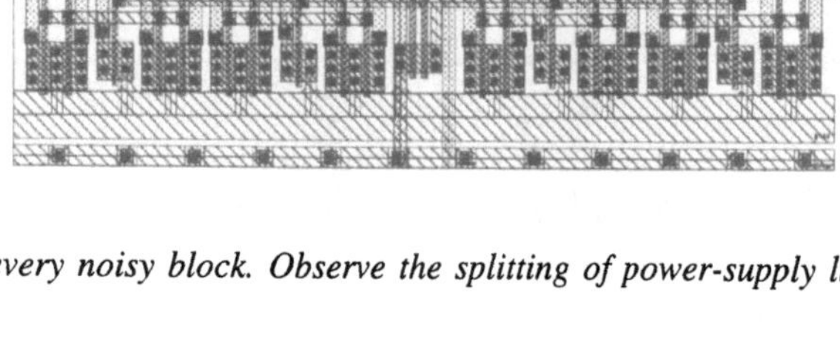

Figure 6.3

Layout of every noisy block. Observe the splitting of power-supply lines and biasing lines.

Any number of blocks between 0 and 32 may be enabled, i.e., 0 to 512 switching buffers. Another 16 noisy blocks driven by an inverted clock signal may be enabled, so the effects of noise compensation can be evaluated. The phase shift between both clocks is also adjustable.

6.1.3. Description of the sensors

The measurements are taken at the output of three different noise sensors. One of them is a noise-acquiring item, which will be called *sensing element*, while the other two noise sensors are comparators, their inputs being connected to an external threshold voltage, and to the output of a sensing element. The characteristics of these three circuits will now be given.

6.1.3.a. Sensing element.

A simple inverting amplifier formed by two transistors is used as a noise sensing element, the dimensions of the PMOS being 50/1.2, and those of the NMOS 20/1.2. A threshold voltage of 2.579 V and a gain of -18.5 are obtained with these dimensions, according to simulation. The transfer characteristic is shown in Figure 6.4. The bias point of the amplifier can be controlled through the gate voltage. In a second version of the sensor both gates can be accessed independently, so as to measure noise as this depends on the bias point of each transistor. The analog section contains certain sensing elements like the one described; in some cases the output of the amplifier is connected to an analog output pad, while in other cases the sensor output is used as an input of one of the comparators. Here the aim is to show that the level of the substrate noise is high enough to produce errors in the comparison, and quantify this drop in resolution.

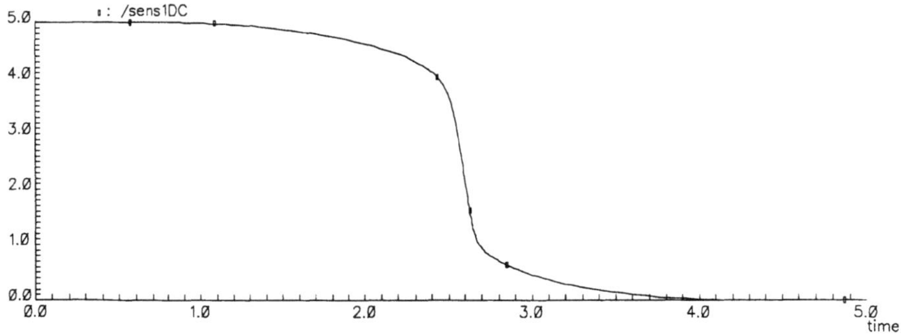

Figure 6.4

Transfer characteristic $V_{ds}(V_{gs})$ of the amplifier used as a sensor.

6.1.3.b. Auto-zeroed comparator.
Apart from measuring the noise coupled through the substrate, we will also prove the influence that such noise has on some common analog circuits. A few experimental studies of this kind are found in the literature: one on the effect of switching noise on ΣΔ comparators [1], and another on the influence of substrate noise on operational amplifiers [2], in which the noise is injected as a square signal between the ground terminals in the circuit.

A typical application where substrate noise is a common source of trouble is A/D converters [3]-[6]. The most noise-sensitive phase in these circuits is the analog voltage comparison. Some sorts of converters –pipeline, flash–, require a high number of comparators, which are usually implemented with auto-zeroed amplifiers because of their size and accuracy [7]. We choose this type of comparator to verify the effect of resolution loss provoked by substrate noise. Auto-zeroed comparators have also been successfully used to reconstruct noise waveforms, detecting the amplitude of a high number of noise samples taken with slight delay increments [8].

The schematic diagram of the circuit implemented is shown in Figure 6.5. Transmission gates in the last inverters are implemented to get a faster response. The non-inverting input is connected to the output of one of the sensors presented before, while the inverting input is fixed to a threshold voltage through an external connection. Simulation results prior to measurement show a minimum resolution of 10 mV.

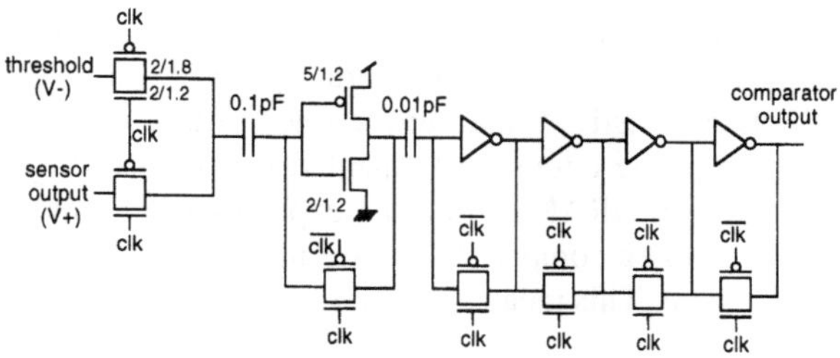

Figure 6.5

Schematic diagram of the comparator used to measure the influence of noise on A/D converters.

Some comparators of this type were implemented in the test IC, to observe the effects of noise produced both by trailing and falling edges at the output of the noisy buffers. Figure 6.6 shows an image of the circuit, in which the capacitors implemented with double polysilicon layers may be seen.

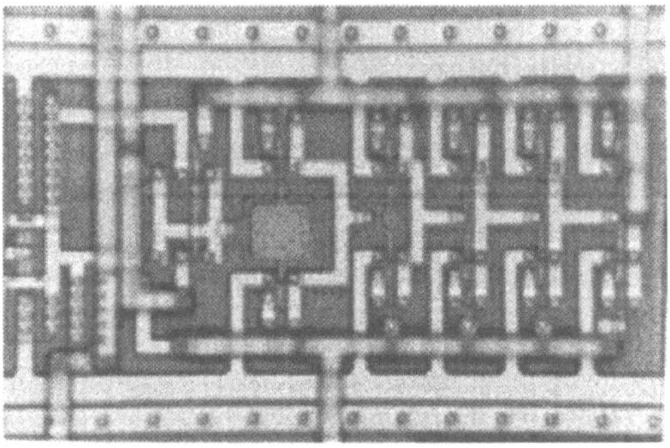

Figure 6.6

Photograph of the auto-zeroed comparator.

6.1.3.c. Sense amplifier.

Another type of circuits that may be sensitive to substrate noise is memory, especially dynamic RAMs. These circuits are usually manufactured with advanced technologies, so they are not integrated together with large digital circuitry. Nevertheless multimedia ICs that include DRAM memories have recently been developed. Experimental measurements are reported in [9] showing dynamic memory refresh time degradation as a consequence of substrate noise. Also in [10] a device simulator is used to show that coupling may produce errors in the reading of a RAM memory cell.

In order to verify the resolution loss in a DRAM reading due to switching noise, and the possibility of error induction, a comparator based on a sense amplifier is implemented, with characteristics imitating the ones of a real dynamic memory cell. The schematic diagram of the circuit implemented is shown in Figure 6.7. The main difference compared with a DRAM is that the capacitors in memory cells have been replaced by the voltages to be compared: an externally controllable threshold, and the output of a sensing element. This means the sense amplifier does not detect a slight

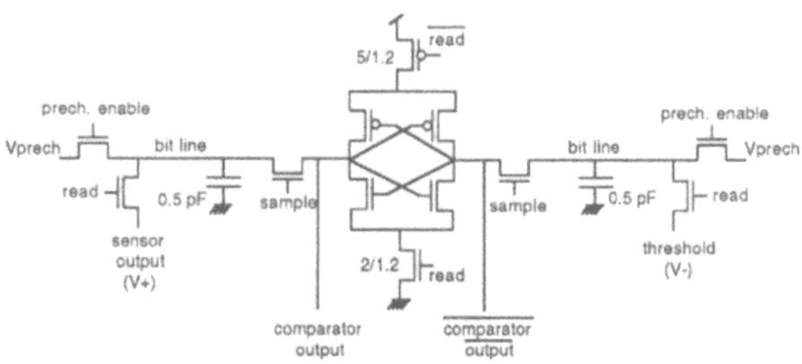

Figure 6.7

Schematic diagram of the comparator used to verify noise influence on the reading of a DRAM cell.

voltage variation due to a charge transfer, as occurs in a DRAM reading, but detects instead the voltage difference between comparator inputs. This means that pass transistors have to be included between the bit lines and the amplifier, so the output is allowed to reach 0 or 5 volts. These pass transistors have the effect of a sampling switch.

The other items, preload transistors and amplifier enabling, are typical in RAM circuits. Figure 6.8 shows the layout of the comparator implemented, where the circuitry generating the control signals is not shown. The 0.5 pF capacitors simulating parasitic capacitances in real memory bit lines stand out.

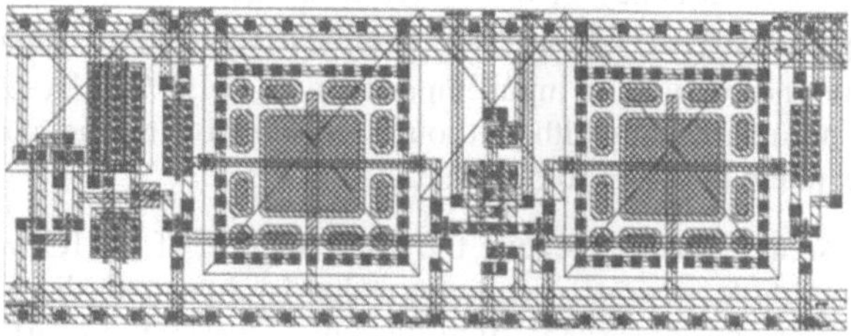

Figure 6.8

Layout of the comparator based on a sense amplifier.

The resolution obtained with this comparator is lower than that of the auto-zeroed amplifier, just 50 mV. This resolution would be enough if the circuit were really to be used in a DRAM: assuming a 2.5 preload voltage and cell capacitances of 40 fF, the voltage difference to detect would be 200 mV. A detail of the comparator showing one of the capacitors can be seen in Figure 6.9, in which layout techniques have been used to get good matching of their values.

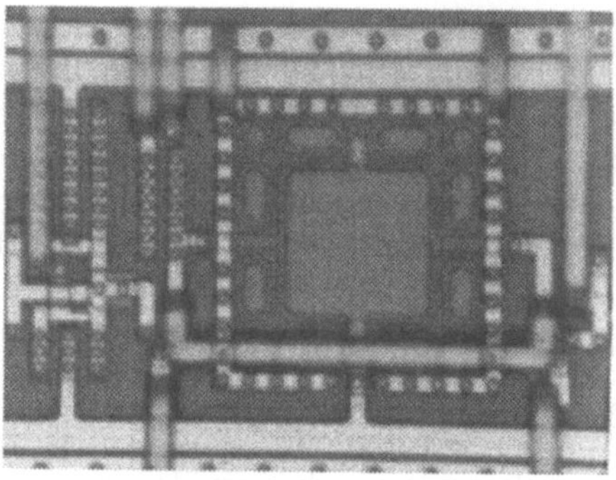

Figure 6.9

Photograph showing a detail of the comparator based on a sense amplifier.

6.2 Noise measurement results

6.2.1. Experimentation environment

In this section the results obtained from noise measurements at the output node of the sensing elements described in 6.1.3.a will be given. The main instruments used are a Tektronix HFS 9003 stimulus generator, which allows pulses with 0.8 ns transient times at a maximum frequency of 650 MHz to be generated, and a Tektronix DSA 602A digital signal analyzer, taking $2 \cdot 10^9$ samples/second. A Tektronix P6205 probe is used to measure noise, with an input impedance of 1MΩ//2 pF, 0.5 ns transient time and 750 MHz nominal bandwidth, although connection to the acquisition module reduces the effective bandwidth to 600 MHz.

A 1 MHz clock signal is used to excite the noisy section of the IC, with 1 ns transient time and 5 V amplitude. A small frequency value is chosen so the noise produced by each transition can be observed, avoiding the overlapping effects of previous transitions. Noise measurements according to the clock frequency were taken, and over 10 MHz the dependence is not monotonous, but peak-to-peak noise depends on additions or cancellations of the noise oscillations resulting from different switchings.

6.2.2. Noise waveforms according to switching direction

Figure 6.10 shows the waveforms of the noise caused by a 0 to 5 V switching of the buffers in 16 noisy blocks, in both kinds of substrates. Biasing is D/A type, and there is a grounded backside contact, as already stated. The sensing element is biased at the maximum gain point, $V_{inDC}=V_{outDC}$. It can be seen that the waveform is approximately the same in both substrates, a double positive-negative peak of considerable amplitude and about 5 ns long, followed by irregular oscillations of smaller amplitude. The peak-to-peak noise level is 115 mV in the P- substrate, and 320 mV in the P+ one, i.e. about three times larger in the second case. The fact that P+ wafers allow easier coupling was already pointed out in chapters 2 and 4, given that propagated noise is not attenuated in the same way. The 10 ns delay appreciated between input switching and commencement of noise is due to the delay chain included in the IC for synchronization purposes, and its value ties in with simulation predictions.

Figure 6.11 shows the waveforms now produced by a falling transient at the outputs of the buffers. The noise levels are somewhat smaller than those produced by a rise transient, but the most important feature to note is the similarity with noise waveforms in Figure 6.10, for a rise transient. This indicates that most of the noise is introduced to the substrate through the digital supply lines, and not through the junctions of the switching transistors. In this second mechanism, the junctions behave like capacitances and produce complementary waveforms in both transients, as seen in chapter 2. Nevertheless, noise at the digital supply lines will have the same shape in

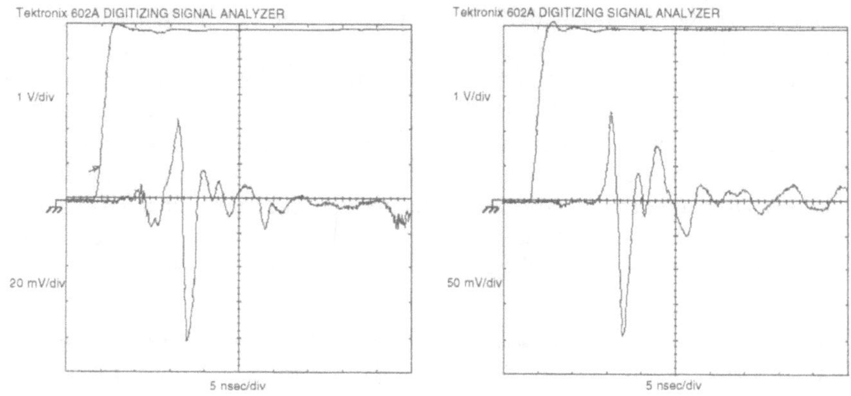

a) *b)*

Figure 6.10

Noise waveforms at the output of the sensor biased at its maximum gain point, for a rise transition of the noisy buffers. a) P- substrate. b) P+ substrate.

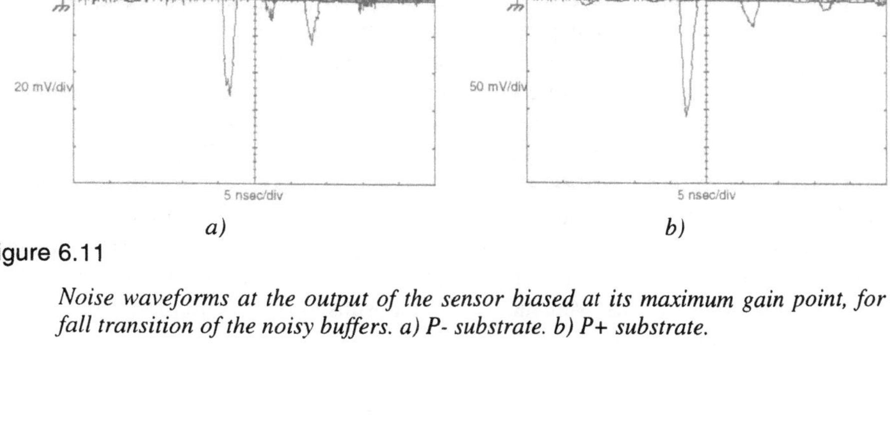

a) *b)*

Figure 6.11

Noise waveforms at the output of the sensor biased at its maximum gain point, for a fall transition of the noisy buffers. a) P- substrate. b) P+ substrate.

both transitions, although symmetric at *GND* and V_{dd} lines. Figure 6.12 shows a diagram of the current flow to charge and discharge the output of a digital gate in both logic transitions. In a 0 to 1 transient at the output of a gate, the load capacitance to *GND* is charged by a current through the PMOS net, while the load capacitance to V_{dd} is discharged through the same net. In a 1 to 0 logic transition, charging and discharging are through the NMOS net, but as we can see, the current through the package inductances takes the same direction, so the voltage waveforms are the same. Apart from this current for charging capacitances, in all CMOS gate transition there is a shortcircuit current between V_{dd} and *GND*. This current is usually much smaller than what is required to charge capacitances [11], but since the noisy buffers are not loaded with any gate in our chip, the shortcircuit current may be dominant (the charge/discharge current is not zero, because the junction capacitances are still present). In any event this shortcircuit current also has the same direction in both transients, and the waveforms at the supply lines will be still the same.

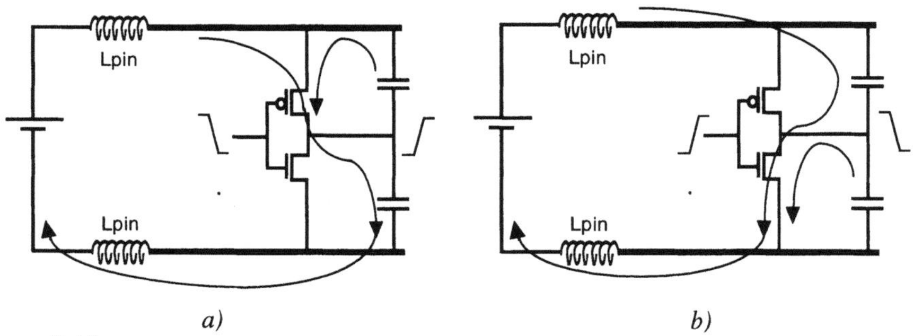

Figure 6.12

Direction of the currents charging and discharging the output loads of a CMOS logic gate. a) In a 0 to 1 logic output transient. b) In a 1 to 0 logic output transient.

The presence of identical waveforms produced by both switching transients makes the noise compensation technique proposed in chapter 5 impossible. In the simulation analysis presented in that chapter it was already stated that, if noise were coupled to the substrate by supply lines, noise compensation would not be feasible. A number of counter-phase switching buffers were implemented in the test IC, with an adjustable delay so as to evaluate the possibility of compensating noise. Figure 6.13 shows the noise produced by both blocks, phase and counter-phase, and it can be seen that waveforms are practically identical. Simultaneous switching of both blocks leads to a noise increase as expected. Some attempts to adjust switching skew were made to attempt to compensate at least one of the peaks, and although there was local compensation, other peaks were added in the subsequent oscillations, resulting in similar or even larger noise.

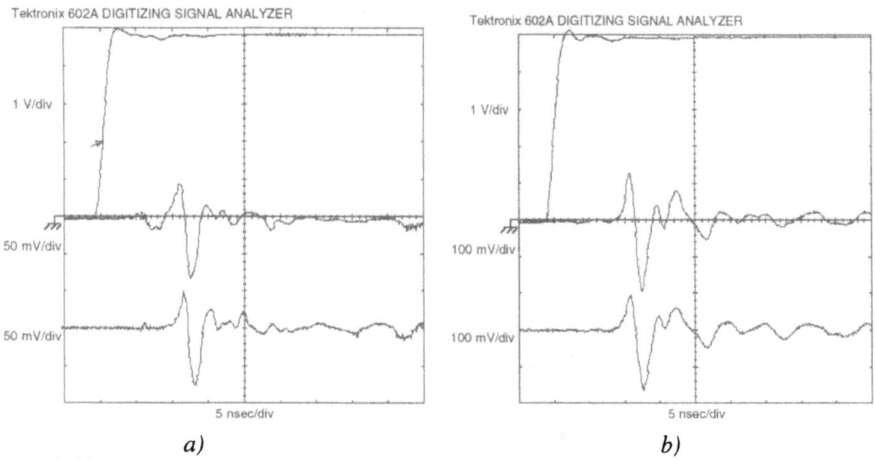

Figure 6.13

Noise waveforms at the output of the sensor caused by counter-phase switching blocks. Waveforms are identical, so noise compensation is not feasible. a) P- substrate. b) P+ substrate.

6.2.3. Noise in function of the sensor bias point

The measurements presented before were done biasing the sensing element at its maximum gain point, V_{inDC}=V_{outDC}=2.555 V. In this section measurements to observe the noise dependence on the sensor bias point and on the transistor state are presented. Figure 6.14 depicts the transfer characteristic measured, which is coincident to the one expected from simulation, represented in Figure 6.4. The noise measured when the sensor is biased at different points of this characteristic is shown in Figure 6.15, for both wafers. It can be seen that noise is clearly smaller when one of the transistors is cut-off. In a P+ substrate it can also be seen that noise increases when placing the sensor at its high gain region. This was already observed in the simulations and measurements in chapter 2. Anyway, this increase is small, and in P- substrates it is not even appreciated. The noise received by the sensor is the addition of contributions of the noise directly coupled from the substrate, plus the contribution of noise at power-supply lines, which will depend on the sensor bias point. If V_{outDC}=2.5 V, the instant noise measured will be:

$$V_{out} = V_{GND} + V_{subs} + \frac{V_{Vdd} - V_{GND}}{2} = V_{subs} + \frac{V_{Vdd} + V_{GND}}{2} \quad Eq. (6.1)$$

that's to say, the average of noise in both power-supply lines plus the noise received

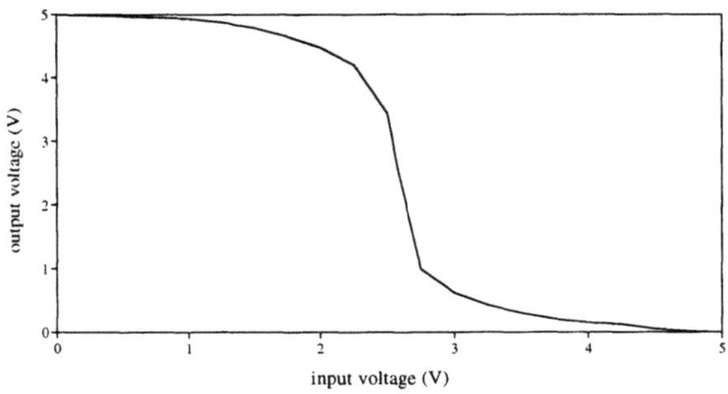

Figure 6.14

Measured transfer curve $V_{ds}(V_{gs})$ of the amplifier used as a sensor.

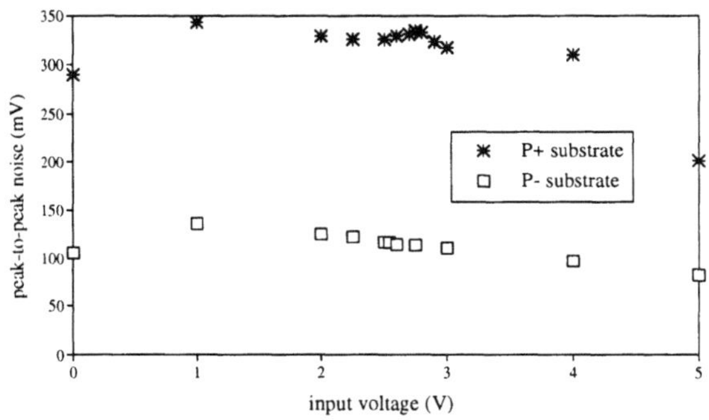

Figure 6.15

Noise measured in function of the bias point of the amplifier used as a sensor.

through the substrate. The gain of the sensor only affects the noise received through the substrate by body-effect. If the noise dependence on the gain is small, this means that most of the noise is received through the supply lines. In P- wafers, noise reaching the sensor through the substrate is smaller than in P+ ones due to attenuation, thus the dependence on the gain is negligible. When the NMOS transistor is cut-off, the noise sensed is the noise present at the analog V_{dd} line, plus the noise acquired through the junctions. On the contrary, if the PMOS is cut off, the noise measured at the output is that of the analog *GND* plus that acquired through the junctions. Figure 6.16 shows the waveforms obtained in these two conditions, in a P+ substrate –very similar waveforms are obtained in P- substrates–. If these two waveforms are averaged, the result obtained is a waveform highly correlated with that shown in Figure 6.10 for a $V_{outDC} \approx 2.5$ V. Thus we can affirm that noise measured at the sensor output is mainly due to the noise at the analog supply lines, especially in highly doped substrates.

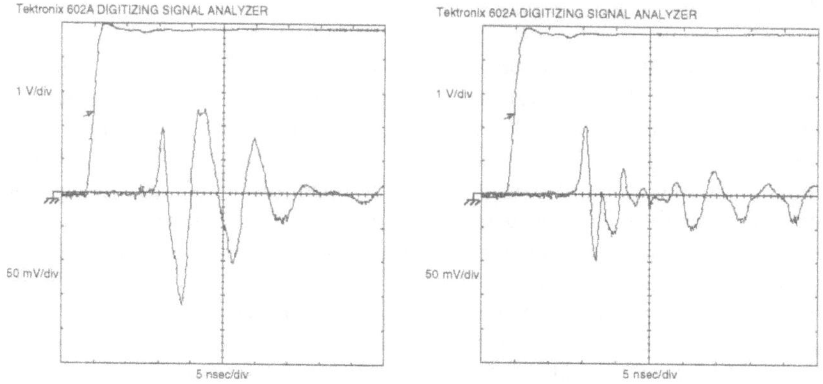

a) *b)*

Figure 6.16

Noise waveforms at the output of the sensor with a 0 V input voltage (NMOS cut off, noise at analog V_{dd}*) (a), and biased with a 5 V input voltage (PMOS cut off, noise at analog* GND*) (b).*

Some of the sensing elements implemented have independent connections to every transistor gate, so the operating region of both devices can be controlled independently. Table 6.II shows the noise measured placing every transistor in its cutoff region, and increasing the gate voltage of the other device so different degrees of channel formation are obtained. As expected, we can observe that as the channel resistance decreases the amount of noise introduced from every supply line increases. The noise detected when both transistors are situated in a high impedance state is mainly acquired through the junctions of the devices.

Table 6.II.

Noise received when varying the channel resistance of each transistor.

V_{gate} NMOS	V_{gate} PMOS	$V_{out\ DC}$	noise (P-)	noise (P+)
0	0	5 V	103 mV	281 mV
0	4.2	4.2 V	52 mV	182 mV
0	5	47 mV	39 mV	128 mV
0	5	47 mV	39 mV	128 mV
2.5	5	110 mV	96 mV	131 mV
5	5	135 mV	133 mV	159 mV

6.2.4. Simulation of the circuit extraction

In order to get a better understanding of the way noise is introduced to the substrate and what the most relevant parameters affecting coupling are, we simulated the full circuit including a model of the P+ substrate, to compare noise levels and waveforms against the ones obtained experimentally. An extraction of the circuit netlist is obtained, including parasitic capacitances and pad protection diodes. P+ bulk is modeled as a single node, and a resistance between every device and this node is included, its value obtained from device dimensions and technological characteristics. A hierarchical circuit extraction is performed in order to facilitate including this substrate model. Furthermore, equivalent resistances between each supply or biasing line and the substrate are also calculated, according to the number and size of the contacts. The pads include substrate contacts which are connected to the corresponding supply lines (analog supplies if this is an analog pad, digital supplies for a digital pad). The total equivalent resistance of the contacts in the digital pads is around that of the total equivalent resistance of the contacts in the digital core, while in the analog domain the equivalent resistance of the contacts in the pads is smaller than that of the contacts in the circuitry. This means that, in P+ substrates, the biasing strategy in the circuit core will be masked by the effect of the pads, as expected from the results in chapter 4.

Every package pin is modeled with the equivalent circuit shown in Figure 4.2, with the inductance values of a DIL package that can be found in [12]. After adjusting the

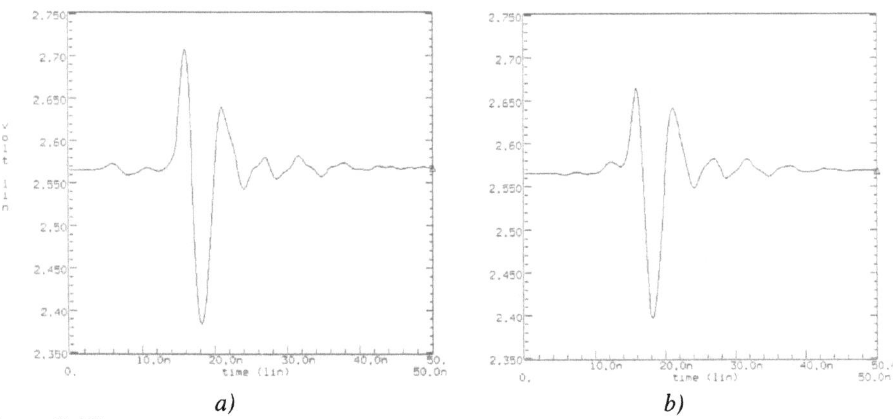

Figure 6.17

Noise waveforms at the output of the sensor biased at its maximum gain point, obtained from the simulation of the extracted circuit, including a P+ substrate model. a) 0 to 5 V transition of the noisy buffers; compare with the measured waveform shown in Figure 6.10.b. b) 5 to 0 V transition of the noisy buffers; compare with the measured waveform shown in Figure 6.11.b.

values of the resistances in the substrate model (the values initially used were too high) the simulation results correlated well with the experimental measurements. Figure 6.17 shows the waveforms in a sensor biased at its maximum gain point in the same conditions as for the experiment presented in 6.2.2. The correlation with the waveforms shown in Figure 6.10.b and 6.11.b can be observed.

Figure 6.18 shows the waveforms obtained by biasing the sensor with 0 and 5 V gate voltages. Again a high correlation is observed with the corresponding experimental waveforms shown in Figure 6.16. Not only are the waveforms coincident, but also the noise levels, as can be seen in Table 6.III. It was also verified that when each of the transistors in the sensing element is cut-off, the noise at its output is coincident with the noise at each supply line, thus confirming that this is the main mechanism through which noise reaches the sensor. As for the path through which noise reaches analog supply lines, a number of simulations were made to check the role of the parasitic elements that might be significant in this path. It was verified that the origin of the disturbance is switching noise at the digital power-supply lines, which is resistively coupled to the substrate through the contacts. This noise is mainly introduced by the contacts in the pads. Every digital pad is surrounded by a biasing ring, thus forming a very low equivalent resistance between ground line and substrate –approximately 15 Ω, in our test IC. Noise is then propagated along the substrate as a single node and couples to the *GND* analog line again resistively, through the substrate contacts. In

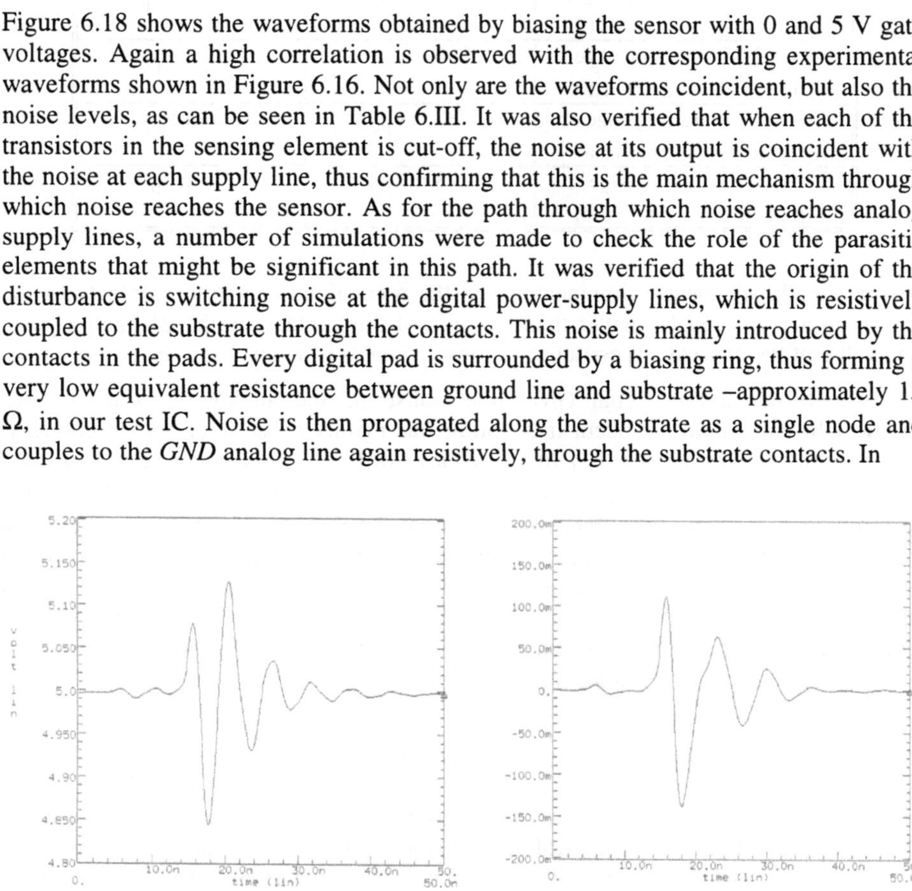

a) *b)*

Figure 6.18

Noise waveforms at the output of the sensor, obtained from the simulation of the extracted circuit including a P+ substrate model. a) Biasing the sensor with 0 V at its input; compare with the measured waveform as shown in Figure 6.16.a. b) Biasing the sensor with 5 V at its input; compare with the measured waveform as shown in Figure 6.16.b.

Table 6.III.

Comparison of noise levels obtained by simulation and with experimental measurements, in the situations represented in Figures 6.16 and 6.17.

Situation:	Simulation	Measurement
V_{in}=2.55 V, trailing edge (fig.6.17.a vs. fig.6.10.b)	321 mV_{pp}	320 mV_{pp}
V_{in}=2.55 V, falling edge (fig.6.17.b vs. fig.6.11.b)	266 mV_{pp}	259 mV_{pp}
V_{in}=0 V, trailing edge (fig.6.18.a vs. fig.6.16.a)	282 mV_{pp}	290 mV_{pp}
V_{in}=5 V, falling edge (fig.6.18.b vs. fig.6.16.b)	248 mV_{pp}	201 mV_{pp}

this case pads are clearly dominant, since the area of the substrate contacts in the analog circuitry is much smaller than the area of the contacts in the 21 analog pads. The value of the equivalent resistance of these contacts is again about ten ohms.

The waveforms at the digital *GND* line, substrate and analog *GND* are almost identical, which would not occur if coupling was mainly capacitive. On the contrary, the waveform at the analog V_{dd} line is clearly different, as can be seen in Figure 6.16.a, because in this case noise is capacitively coupled through the wells and the parasitic capacitances of the line. In conclusion, the number of substrate contacts in the pads and the fact of their being connected to digital and analog supplies allow noise easy access to the sensitive circuits in P+ substrates, regardless of the measures taken in the layout of the circuit core. The only measures that could be effective in this type of substrate are approaches involving circuit topology –differential circuitry– but even so, high power-supply rejection ratios should be achieved.

6.2.5. Noise according to the activity of the digital section

Figure 6.19 depicts the amount of noise measured according to the number of noisy blocks enabled. Behavior is identical in both kinds of wafers, a sustained increase with the number of switching gates. The relation is not linear, but there is a marked tendency towards a saturation that would be reached for an extremely high number of switching gates. One should remember that 32 enabled blocks means 512 buffers synchronously switching.

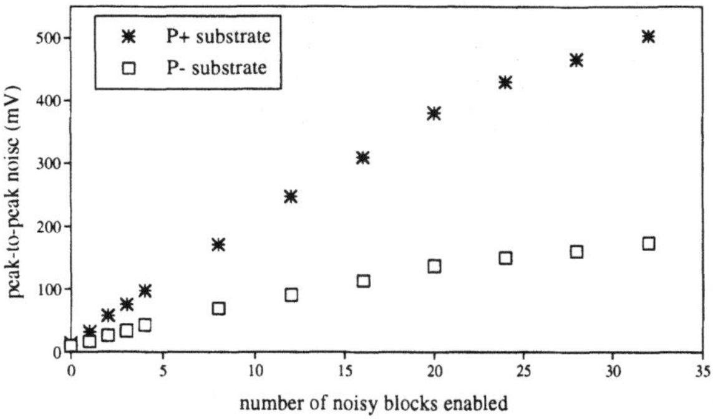

Figure 6.19

Noise measured according to the number of blocks switching.

6.2.6. Noise according to the biasing strategy

In this section the results of the measurements taken with different biasing strategies are presented. One should take into account that the contacts that bias N-wells and substrate in every domain –digital, analog, guard ring– can be connected to the different supplies available through low resistive switches. The same terminology as defined in chapter 4 will be used in the presentation of the results.

6.2.6.a. Biasing strategy in P+ wafers

The analysis in chapter 4 concluded that in epi wafers disturbance is easily coupled to supply lines through pads, making the biasing strategy at the circuit core immaterial. The results were corroborated in the simulations of the extracted circuit, so a significant influence of the biasing strategy was not expected in the measurements. Figure 6.20 shows the results of the measurements in this substrate type, confirming what was expected: very similar noise levels, with no significant differences between the biasing approaches. The waveforms obtained are also almost identical in all cases. When the guard ring was biased, significant variations were not obtained either, even when connected to a dedicated ground line.

The ICs are also biased with a backside contact double-wired to the ground pins exclusively dedicated to bias substrate. This fact was taken advantage of to progressively eliminate both wires in one of the samples, and verify what the effect of the grounded backplane is. The result was that when one of the wire-bondings was eliminated, noise increased from 320 mV to 325 mV, and when the second bonding was eliminated, noise reached 350 mV. The presence of a biased backplane therefore only makes a slight contribution to noise reduction, as expected from the simulation results in chapter 4.

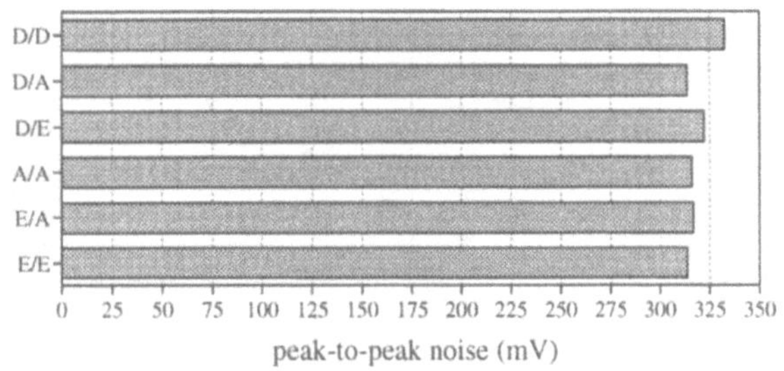

Figure 6.20

Noise measured according to the biasing strategy. P+ substrate.

6.2.6.b. Biasing strategy in P- wafers

Significant influence of the biasing strategy in P- substrates was initially expected, because even though switching noise at digital supply lines will be similar to what is found in a P+ substrate, the disturbance is now attenuated when propagated along the substrate. This means that the noise received by the contacts in the analog pads is small, and the contacts in the circuit core play a more important role. Nevertheless, it can be seen in Figure 6.21 that the noise levels obtained are highly uniform, and in any event the dependence on biasing does not have the logic expected from the results in chapter 4. Strategies such as A/A or A/D being better than D/A or E/A is not understandable. Similar results –characterized by noise uniformity and a senseless relationship– are obtained if the activity of the noisy section is reduced, or if the guard ring is grounded.

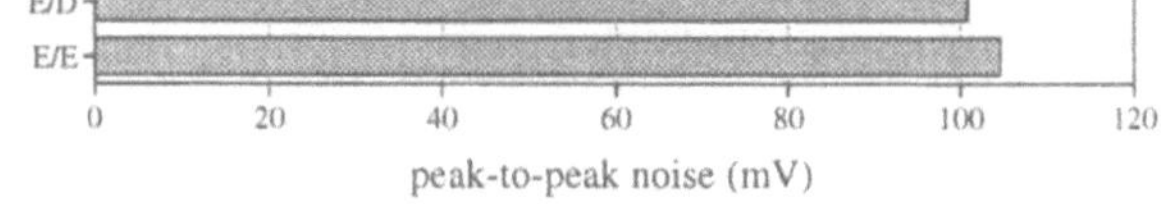

Figure 6.21

Noise measured according to the biasing strategy. P- substrate, 16 noisy blocks enabled.

It is thus risky to draw any conclusions about biasing strategies from the results obtained. On the contrary, the dominant noise propagation mechanism would seem to have little to do with the circuitry in the core, and the differences between the values obtained are mainly due to second-order effects, like resonance, coupling between lines, or the same coupling across the junctions of the devices.

The uniform noise levels measured, and the identical appearance of the noise waveforms in P- substrates and P+ substrates, regardless of the biasing strategy, means that the real coupling mechanism is the same in both cases. Therefore, in the lightly doped substrate too, disturbance is injected to the substrate from the digital supply lines, and then reaches the analog supply through the pads. In P- substrates noise is attenuated in its propagation, but not as much as expected. This attenuation is a consequence of the substrate contacts, that provide a low impedance return path for the disturbance, as was seen in chapter 2. The parasitics of the pins increase their impedance, so there is no good return path, and noise is not attenuated with the expected intensity.

Consequently, when applying noise measures to reduce substrate noise, the primary concern should be to minimize noise at the supply lines and coupling to the substrate at the chip periphery. If the influence of the pad structures is minimized, substrate coupling will be caused at the circuit core, and the strategies discussed in chapter 4 will apply. If the dominant effect of the chip periphery cannot be avoided, the biasing strategies discussed in the aforementioned chapter can also be applied to the contacts in the pads.

6.3 Results of the measurements on the comparators

6.3.1. Effect of noise on an auto-zeroed comparator

The aim of the measurements on the comparators is to show that substrate coupled noise can produce errors in comparisons and a drop in resolution. We started by determining the resolution of the comparators when the noisy section is disabled, measuring the maximum difference of input voltages that produces incorrect comparisons. The resolution obtained is much worse than what was expected from simulation, due to the noise produced by the input clock. The sampling signals have an adjustable delay in respect of the input clock switching (and noisy buffer switching). Effective resolution thus depends on the delay of the sampling signal. In the worst case, i.e. to guarantee a correct comparison for any delay, a resolution of 50 mV is measured if $V- > V+$, and of 200 mV when $V+ > V-$.

It has been shown that noise produced by digital section's activity can produce comparison errors, when noise is produced either by trailing or falling edges, when the threshold voltage is over or under the output voltage of the sensing element, and both in P+ and P- substrates. As an example, Figure 6.22.a shows the comparator output when *V+*=2.52 V and *V-*=2.56 V, obtaining 0 V as a result of the successive comparisons. Figure 6.22.b shows the output with exactly the same comparison conditions, but now enabling the noisy section, obtaining 5 V as a result in this case. Note that here the error is produced by a falling edge at the output of the noisy buffers.

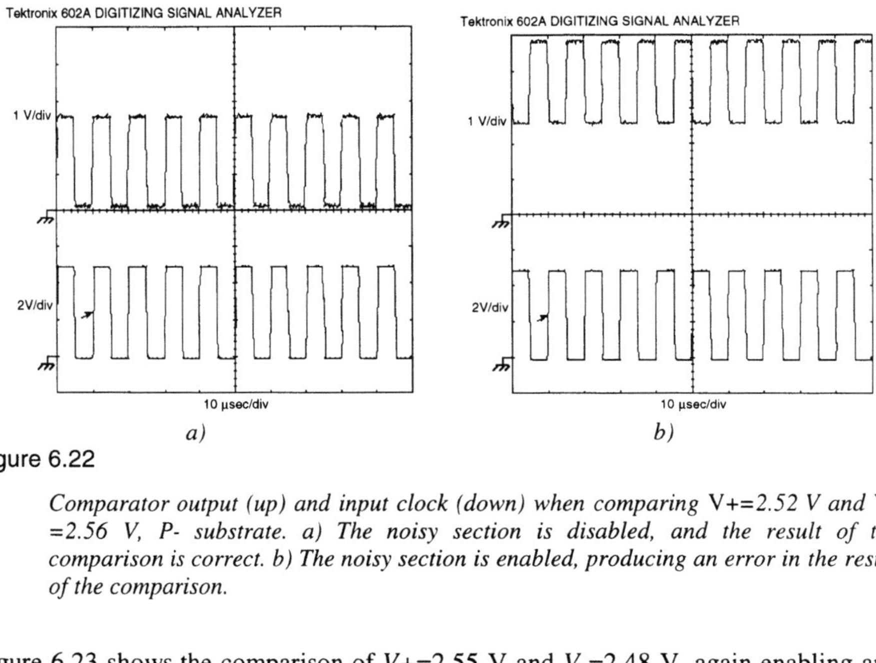

Figure 6.22

Comparator output (up) and input clock (down) when comparing V+=2.52 *V and* V-=2.56 *V, P- substrate. a) The noisy section is disabled, and the result of the comparison is correct. b) The noisy section is enabled, producing an error in the result of the comparison.*

Figure 6.23 shows the comparison of *V+*=2.55 V and *V-*=2.48 V, again enabling and disabling the noisy section. In this case, the disturbance is produced by a 0 to 5 V transient at the output of the noisy buffers. Note that now the noise does not produce 100% errors, but only roughly 50%. In fact, when the threshold is increased from a value under V+ to a value over V+, there is no sudden jump from 100% comparisons giving 0 V as a result to 100% comparisons giving 5 V, there being a metastable zone, as illustrated in Figure 6.24. This metastability is due to random noise affecting both input voltage levels and synchronism. The width of this metastable zone will be larger when sampling a signal with high slopes. In our measurements the width of this metastable zone varies according to the time of sampling, but it is always under 50 mV.

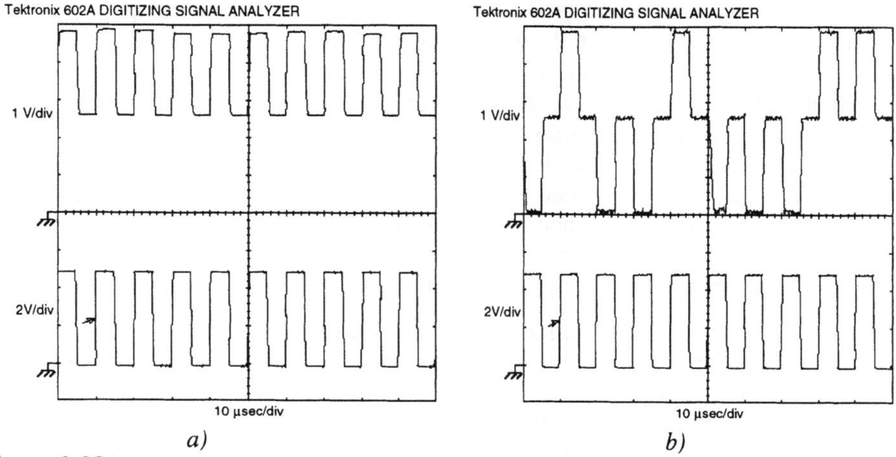

Figure 6.23

Comparator output (up) and input clock (down) when comparing V+=*2.55 V and* V-=*2.48 V, P+ substrate. a) The noisy section is disabled, and the result of the comparison is correct. b) The noisy section is enabled, producing some errors in the result of the comparison.*

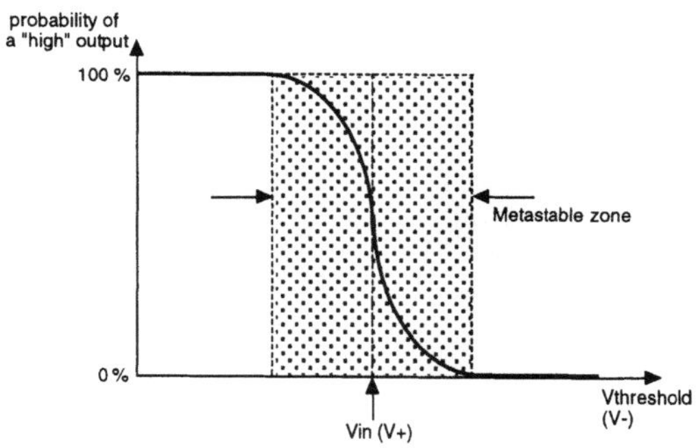

Figure 6.24

Random noise produces a metastable zone in the comparison, and so the probability of having a certain result at the output depends on the difference in voltages at the inputs of the comparator.

If the average of the random noise that produces this metastable zone is zero, then the threshold voltage *V-* producing a 50% probability of a high output is the voltage equal to the positive input of the comparator, *V+*. By determining the value of this threshold voltage according to the delay of the sampling time we may establish the loss of resolution in the comparator produced by switching noise, according to the skew between the sampling time and the switching of the noisy blocks. In fact, this is equivalent to reconstructing the noise waveform by taking samples with small time increments. This technique was already proposed in [13] to reconstruct the noise waveform at the *GND* line in a digital circuit, and was later used in [8] to reconstruct the waveform of substrate coupled noise, also using auto-zeroed comparators. The same authors had previously used this comparator to measure noise according to the activity of the digital section and according to the frequency, although they had to obtain a statistical evaluation of the noise level because they did not synchronize the sampling with the switching of the digital gates [14].

Figure 6.25 depicts the threshold according to the time skew, when 32 and 256 noisy buffers are enabled. The threshold voltage that provides a 50% probability of a given voltage at the output was ascertained by averaging a large number of acquisitions of the output signal, and adjusting the threshold until an average output of 2.5 V was obtained. The waveform reconstructed is not equal to the ones shown in the former section, which may be attributed to the effect of noise on the first stages of the comparator. It can be seen that the effect of noise is extremely large, almost a 1.6 V difference between maximum and minimum threshold values in the worst case. This is much larger than the 320 mV measured directly at the output of the sensing element, which means that the noise detected not only comes from the amplifier generating the *V+* voltage, but from the comparator itself. This is favored by the single-ended topology implemented.

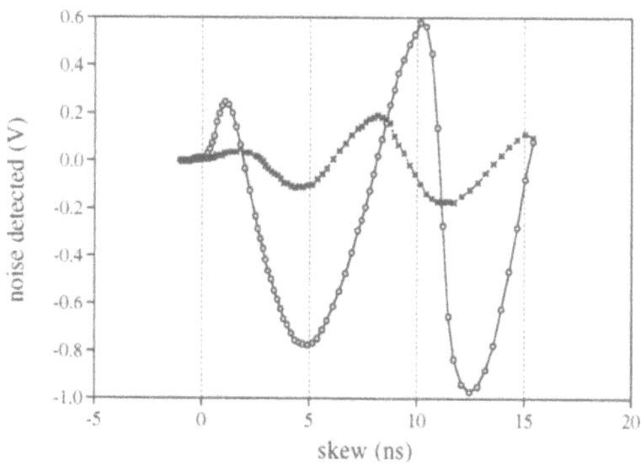

Figure 6.25

Effect of digital noise according to the time skew between the comparator sampling, and the digital buffer switching, for 32 buffers enabled (dashes) and for 256 buffers enabled (unbroken line).

Figure 6.26 shows the reconstruction of the noise detected according to the time delay between the sampling and the switching of the noisy blocks, for 256 enabled buffers, both for the comparators manufactured on a P+ epi and on P- wafers. Again the results with a P+ epi wafer are worse than with P- substrates, although the noise levels in this last wafer continue to be unacceptable. Thus using P- wafers instead of P+ cannot be considered the only solution to substrate coupling problems, but merely an elementary measure.

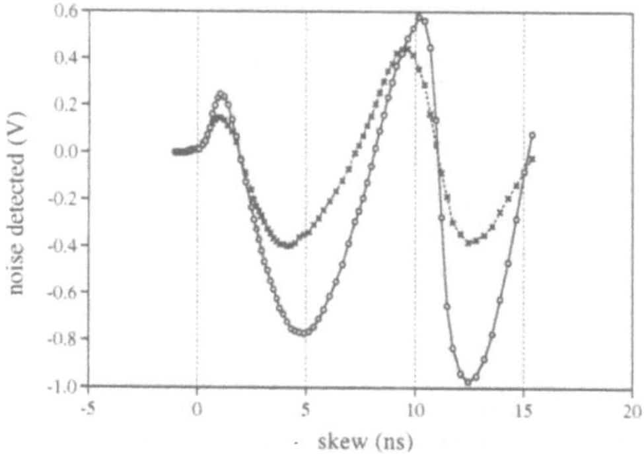

Figure 6.26

Determination of the noise level detected according to the time skew between the sampling time and the switching of noisy gates, in a P- uniform wafer (dashes), and in a P+ epi wafer (unbroken line).

6.3.2. The effect of noise on a sense amplifier

We also started by determining the effective resolution of this second type of comparator, when the noisy section is disabled. The input clock again generates a noise level high enough to make effective resolution worse than expected, again depending on the sampling time. Selecting a preload voltage of 2.4 V, a difference between input voltages of at least 90 mV is needed to guarantee a correct comparison at any sampling time, in case $V- > V+$, and a difference of 210 mV in case $V+ > V-$.

It is seen that when the noisy section is enabled, errors are produced in the comparison, caused both by rising and falling edges at the output of the noisy gates, selecting a threshold voltage above or under the output of the sensing element, and in both P+ and P- wafers. As an example, Figure 6.27 shows the comparison of $V+=2.58$ V and $V-=2.8$, when preload voltage is 2.4 V. Both output waveforms are obtained in the same conditions, except for enabling 256 noisy buffers in the measurement of the right, which produces an error in the comparison. Similarly, Figure 6.28 shows an error in the complementary comparison, now with $V+=2.63$ V and $V-=2.45$ V.

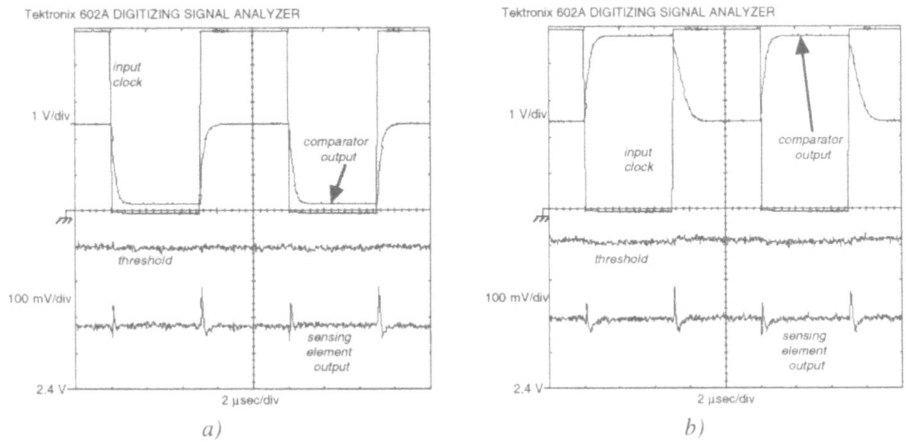

Figure 6.27

Input clock and comparator output (up) and input voltages V+=*2.58 V and* V-=*2.8 V (down). a) The noisy section is disabled, and the result of the comparison is correct. b) The noisy section is enabled, producing errors in the result of the comparison.*

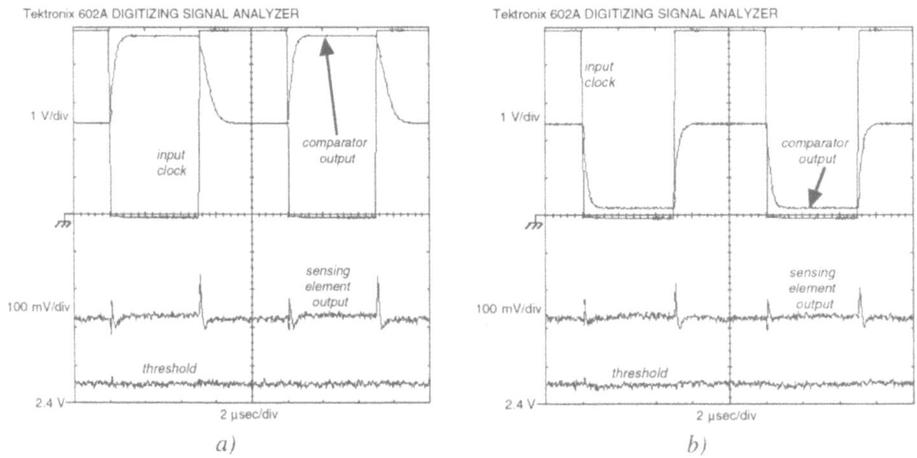

Figure 6.28

Input clock and comparator output (up) and input voltages V+=*2.63 V and* V-=*2.45 V (down). a) The noisy section is disabled, and the result of the comparison is correct. b) The noisy section is enabled, producing errors in the result of the comparison.*

The resolution loss produced when noisy gates are enabled is much larger in this comparator than in the auto-zeroed one. Enabling 16 noisy blocks and measuring the

ICs on P+ substrates, a difference between input voltages of at least 860 mV is needed to guarantee a correct comparison at any sampling time for *V+* > *V-*, while in the *V-* > *V+* case, there is a sampling time at which the comparison is always erroneous. Compared with the auto-zeroed comparator, more control signals appear in the sense amplifier comparator. Synchronism of these signals is important to get good performance, and noise may affect it. Furthermore, noise at the analog supply lines influences the circuit both through the sense amplifier and –in our particular implementation– through the bit line capacitances. The use of this comparator to read the value of a dynamic RAM cell would not be feasible in the conditions of this IC.

References

[1] T. Blalack, B.A. Wooley, "The Effects of Switching Noise on an Oversampling A/D Converter", *Proc. IEEE International Solid-State Circuits Conference*, pp 200-201, 1995.

[2] J. Catrysse, "Measured Distortion of the Output-Waveform of an Integrated OPAMP Due to Substrate Noise", *IEEE Trans. Electromagnetic Compatibility*, vol. 37, no. 2, pp. 310-312, May 1995.

[3] B.P. Brandt, B.A. Wooley, "A 50-MHz Multibit Sigma-Delta Modulator for 12-bit 2-MHz A/D Conversion", *IEEE J. Solid-State Circuits*, vol. 26, no. 12, pp. 1746-1756, December 1991.

[4] J.W. Fattarusso, S. Kiriaki, M. de Wit, G. Warmar, "Self-Calibration Techniques for a Second-Order Multibit Sigma-Delta Modulator", *IEEE J. Solid-State Circuits*, vol. 28, no. 12, pp. 1216-1223, December 1993.

[5] D.B. Ribner, R.D. Baertsch, S.L. Garverick, D.T. McGrath, J.E. Krisciunas, T. Fujii, "A Third-Order Multistage Sigma-Delta Modulator with Reduced Sensitivity to Nonidealities", *IEEE J. Solid-State Circuits*, vol. 26, no. 12, pp. 1764-1774, December 1991.

[6] J. Eklund, R. Arvidson, "A Multiple Sampling, Single A/D Conversion Technique for I/Q Demodulation in CMOS", *IEEE J. Solid-State Circuits*, vol. 31, no. 12, pp. 1987-1994, December 1996.

[7] *Principles of CMOS VLSI Design: A Systems Perspective*, 2nd edition, Neil H.E. Weste, Kamran Eshraghian, Addison-Wesley, 1992.

[8] K. Makie-Fukuda, T. Anbo, T. Tsukada, T. Matsuura, M. Hotta, "Voltage-Comparator-Based Measurement of Equivalently Sampled Substrate Noise Waveforms in Mixed-Signal Integrated Circuits", *IEEE J. Solid-State Circuits*, vol. 31, no. 5, pp. 726-731, May 1996.

[9] T. Tsuruda, M. Kobayashi, M. Tsukude, T. Yamagata, K. Arimoto, M. Yamada, "High-Speed/High Bandwidth Design Methodologies for On-Chip DRAM Core Multimedia System LSI's", *IEEE J. Solid-State Circuits*, vol. 32, no. 3, pp. 477-482, March 1997.

[10] X. Aragonès, F. Moll, M. Roca, A. Rubio, "Analysis and Modelling of Parasitic Substrate Coupling in CMOS Circuits", *IEE Proceedings-Circuits Devices & Systems*, vol. 142, no. 5, pp. 307-312, October 1995.

[11] D.J. Allstot, S. Chee, S. Kiaei, M. Shrivastawa, "Folded Source-Coupled Logic vs. CMOS Static Logic for Low-Noise Mixed-Signal ICs", *IEEE Trans. Circuits and Systems I: Fundamental Theory and Applications*, vol. 40, no. 9, pp. 553-563, September 1993.

[12] *Influència de les Interconnexions en el Disseny Microelectrònic*, Francesc Moll, PhD Thesis, DEE-UPC, Barcelona, 1995.

[13] P. Larsson, C. Svensson, "Measuring High-Bandwidth Signals in CMOS Circuits", *Electronics Letters*, vol. 29, no. 20, pp. 1761-1762, 30th September 1993.

[14] K. Makie-Fukuda, T. Kikuchi, T. Matsuura, M. Hotta, "Measurement of Digital Noise in Mixed-Signal Integrated Circuits", *IEEE J. Solid-State Circuits*, vol. 30, no. 2, pp. 87-92, February 1995.

7 The cause of switching noise

This chapter will attempt to explain the cause of switching noise: its generation, transmission and the effects produced in analog and digital circuits. The analytical models presented in literature and its main parameters will be discussed.

7.1 Mechanisms generating switching noise

Switching noise is also known as **delta-I noise** due to the mechanism generating this type of noise: the voltage spikes that appear across parasitic inductors of IC supply pins when a current pulse flows through these. The existence of such parasitic inductors is due to the package pins and bonding wires and to the whole power distribution network. The value of the dI/dt noise peaks is proportional to the inductance multiplied by the time derivative of the power supply current pulse waveform.

7.1.1 CMOS integrated circuits

The basic structure of a CMOS gate is presented in Figure 7.1. Two complementary transistor networks make up each CMOS gate: the PMOS network connects the output node to the positive power supply node and the NMOS. Both networks work complementarily. The input values and the function of the gate determine which of the two networks has to be on with the other thus being off.

To understand the cause of switching noise properly it is also necessary to include input and output capacitances in the circuit model. The former are due to the gate overlapping over the drain and source regions beside the channel. These capacitances are small compared with the gate capacitance which forms the main contribution to input capacitance. The intrinsic output capacitance is mainly due to the diffusion capacitance of the drain of the transistors connected to the output node. A further contribution to the output capacitance is made by the parasitic capacitance between the metal lines and the substrate or the well. This may be the dominating load capacitance in today's submicron integrated circuits. All these capacitances are modeled by two capacitors that are connected between the output of the gate and the positive and negative power supply nodes, respectively (C_p and C_n in Figure 7.1).

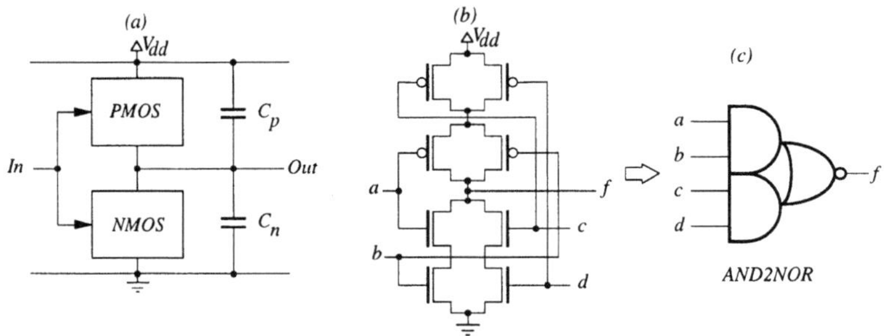

Figure 7.1

(a) Block diagram of a general CMOS gate. (b) Static CMOS AND2NOR gate circuit schematic. (c) AND2NOR gate logic symbol.

When the inputs change in such a way that the output has also to change from high to low or vice-versa some of the output capacitances must be discharged and some of them must be charged, producing a transient power supply current pulse. The amplitude and width of this current pulse determine the amount of switching noise generated. Figure 7.3 shows the simulation waveforms obtained in a high to low output transition in the CMOS inverter shown in Figure 7.2. The current pulse flows through the two package supply terminals and the noise pulse appears in both internal positive power supply node (V_{DD}) and internal ground node (*Gnd*). The result is a transient reduction of the dynamic range of the power supply voltage that may have effects on other quiet circuits and on the switching circuit itself. When several gates in a circuit switch simultaneously the switching current pulses are accumulated. Switching noise thus increases with the number of simultaneous switchings and is called Simultaneous Switching Noise (SSN).

The shape of the current pulses in the high to low output transition may be different from the shape in the low to high transition because the current flows through different transistors and charges and discharges different capacitors. However, if the logic gate is designed to have symmetrical switching times the two output transitions should be expected to demand very similar current pulses. Another reason for variations in the current pulse shape is the input combination that forces the output transition.

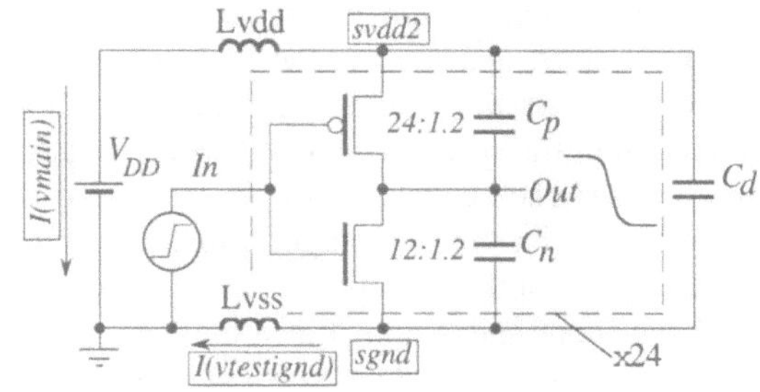

Figure 7.2

CMOS inverters loaded with a fanout of 4 used for simulation. Lvss=Lvdd=1nH. The on-chip decoupling capacitor is C_d=1pF. The circuit consists of 24 simultaneously switching inverters connected to the same power supply nodes (svdd2 and sgnd).

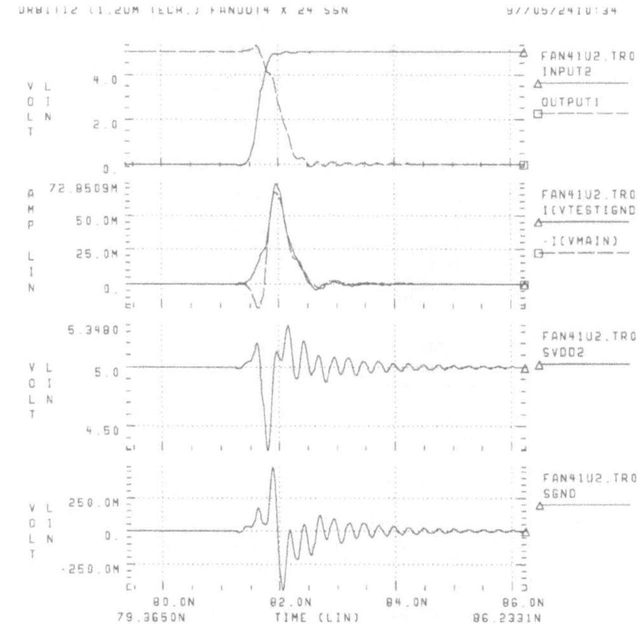

Figure 7.3

Simulation results for the switching currents and noise generated in the internal power supply nodes (agnd and svdd2) for 24 simultaneously switching inverters designed with a 1μm CMOS technology. The names of the waveforms correspond to the labels in Figure 7.2.

There are several important conclusions to be drawn from Figure 7.3 connected with the mechanism generating switching noise. Firstly, it can be seen that the current pulse flowing through the inductor labeled *Lvss* is slightly different from the current pulse flowing through *Lvdd*. There is a non-symmetrical current loop between the generator, the gates of the MOS transistors, and *Lvss* in the high to low transition because the generator is referred to the external ground node. The same loop appears through *Lvdd* in the low to high output transition. This is the same situation as is found in the output drivers because the switched capacitances are situated off-chip and the return path for the current is different for the two output transitions (high to low and low to high). In these cases the noise spike is different in the two internal *Vdd* and *Gnd* nodes. However, if all the input and output transitions are internal, the loop of the switching current pulses comprises both *Lvss* and *Lvdd* and the voltage noise spikes are identical but with reversed sign in the internal *Vdd* and *Gnd* nodes (assuming that the two inductances are equal). Secondly, some oscillations can be observed after the main voltage spike at the power lines. These oscillations are due to the RLC circuit formed by the package inductance and the package and circuit capacitance and resistance. This RLC circuit is complex and can have several resonant frequencies. The damping of such a resonant circuit must be high enough to prevent the oscillations produced by one switching event being added to the next switching event because this would successively increase the noise pulse amplitude.

The switching noise generated by a gate or an output driver depends on the output load. Output MOS transistors can be in two working zones while switching: the ohmic or linear zone and saturation, depending on the ratio between the gate to source voltage (V_{GS}) and the drain to source voltage (V_{DS}). The voltage V_{GS} depends on the input and the voltage V_{DS} depends on the output. The output falling (or rising) time depends on the current driven by the output transistor(s) and the load capacitance.

Switching transition at the output of a CMOS gate can be divided into the different sections illustrated in Figure 7.4 for a CMOS inverter. The figure shows an output falling transition (high to low). While the input voltage minus the threshold voltage is greater than the output voltage the NMOS is in a saturation state. Once V_{GS}-V_{TN} falls bellow V_{DS} the NMOS enters the ohmic zone. For a fraction of the transition the output load discharging current depends on the square or the input voltage. For the rest of the transition the current also depends on the output voltage waveform and is always smaller than the saturation current. Good design will mean that the NMOS will be in a saturation state for almost all the transition time. The load capacitance should be greater the longer the output falling time. If the time taken by the output to fall from V_{DD} to V_{DD}–V_{TN} is greater than the input rising time from 0 to V_{DD} all the falling transition occurs with the NMOS in a saturation state. This is the typical situation in VLSI circuits where the fanout[1] is optimized for a fast transition for all the gates.

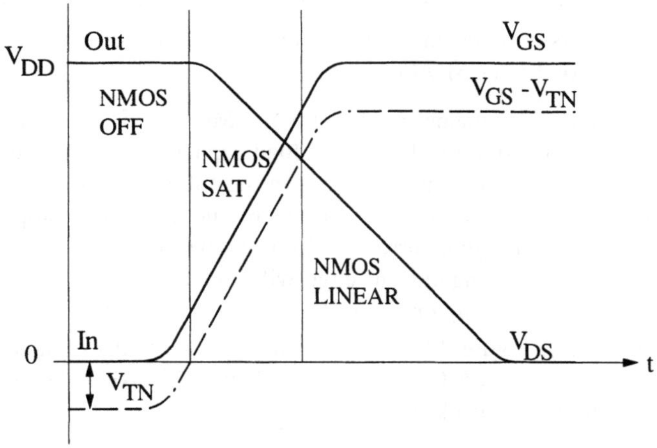

Figure 7.4

Output and input waveforms for a CMOS inverter indicating the NMOS working regions for a high to low output transition.

The influence of the load in CMOS circuit switching was studied in [1], where it is shown that above a critical value for the load capacitance (C_{crit}) noise does not increase with output load because the maximum current is limited by the saturation current for $V_{GS} = V_{DD} - V_{TN}$, and its slope is limited by the input transition time. The value of the critical load capacitance can be calculated as:

$$C_{crit} = C_0 \left[\frac{M}{n} - \sqrt{\frac{M^2}{n^2} + \frac{2M}{n}} \right]^2 \qquad Eq.\ (7.1)$$

where

$$M = \frac{t_r}{2V_{DD}\beta L} \quad C_0 = \frac{\beta t_r}{6V_{DD}V_T}[V_{DD} - V_T]^3 \qquad Eq.\ (7.2)$$

[1]The fanout of a gate is defined as the ratio between the output load capacitance and the input capacitance. One inverter loaded with four identical inverters to it will have a fanout of four.

In these expressions n is the number of inverters or drivers that switch simultaneously, t_r is the rising (falling) time, V_{DD} is the supply voltage, β is the NMOS (PMOS) gain, L the channel length and V_T the threshold voltage of the NMOS (PMOS), where the expressions in brackets correspond to an output rising transition and the others to an output falling transition for a CMOS inverter.

In the above equation long channel models for the MOS devices are used. For short channel devices the expression for the critical load capacitance would be different but the physical behavior would remain the same. In the following case the CMOS gates will be considered to be sufficiently loaded to overcome the critical output load value. The load charging or discharging current will always be assumed to be the maximum saturation current and the rising current edge will take as long as the input rising or falling time from V_{TN} to V_{DD} or from $V_{DD} - V_{TP}$ to 0, for NMOS and PMOS transistors, respectively. Figure 7.5 shows the simulation results obtained by varying the load capacitance of the CMOS inverter of Figure 7.2. Also shown is the noise generated by the circuit for a package parasitic inductor value of $L_{eff} = 8$nH.

From the figure it can be observed that for loads below 0.5 pF the current pulses fail to reach the maximum value. Beyond 0.5 pF the rising edge of the current pulse is always the same and the falling current edge slope drops as the output load increases. This is the reason why the negative noise spike at internal V_{DD} node (and also the positive spike at internal *Gnd* node) is almost constant although the output load is increased. The level and the frequency of the oscillation vary because the output load has an influence on the RLC resonance of the circuit formed by the package inductance and the IC.

When the load capacitance is greater than the critical capacitance the noise generated depends only on the amplitude and the duration of the rising edge of the current pulse. Current pulse amplitude depends on the transistor size and current pulse rising time depends on the input rising or falling time. If the load capacitance is smaller than the critical capacitance current pulse shape also depends on the output voltage and the short circuit current must be also considered. In this last case, for an increasing load capacitance the output transition is slower and the noise generated is smaller [2]. This behavior is observed only if the fanout is small and if the output MOS transistor works in the ohmic region for the main part of the transition.

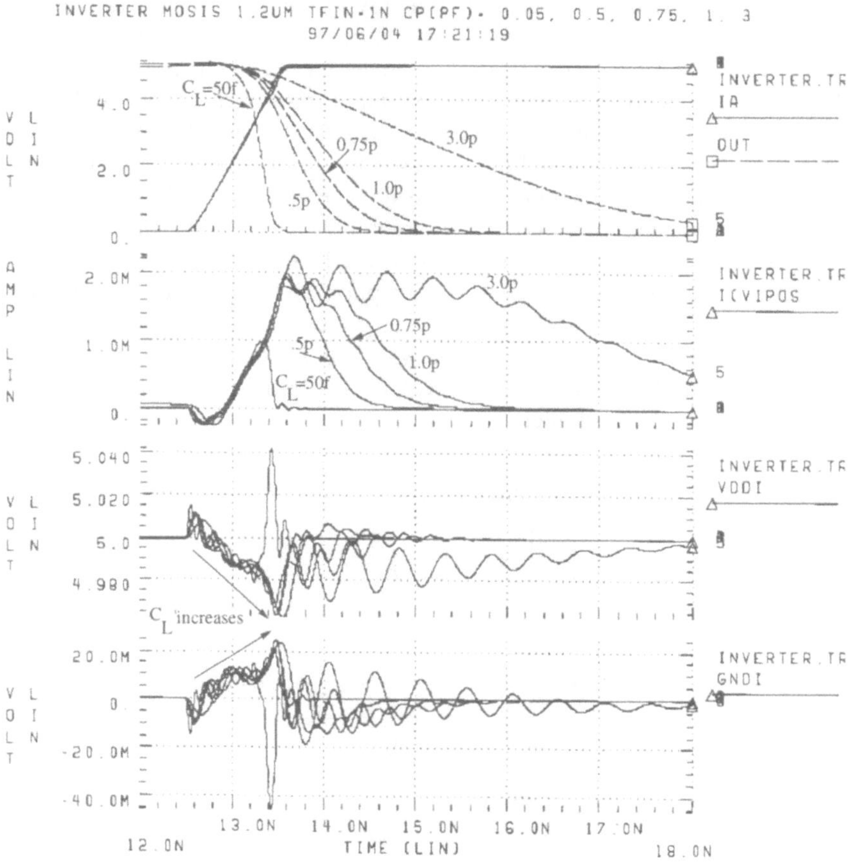

Figure 7.5

Simulated waveforms for the inverter of Figure 7.2 loaded with different capacitor values. The upper panels are used for the input and output waveforms. The second panel shows the power supply current. The two lower panels show the internal V_{DD} and Gnd noise waveforms.

7.1.2 BiCMOS logic circuits

BiCMOS logic gates combine CMOS and bipolar devices to perform a logic function. A complementary CMOS gate performs the function itself and bipolar transistors are added at the output stage of the gate to increase its driving capability and to make the gate faster. It is interesting to analyze the switching noise generated by the BiCMOS gates because they are used to fast- switch highly loaded signals such as the clock or the internal buses. Figure 7.6 shows the structure of a standard BiCMOS inverter. There are other BiCMOS topological alternatives but all of them are based on this standard structure [3][4]. The circuit has two input transistors, a PMOS (M1) and a NMOS (M2), each followed by a bipolar transistor (Q1 and Q2) in a totem pole configuration. Transistors M3 and M4 are used to cut off one of the BJTs when the other is charging or discharging the output load. When the input falls, M1 switches on and drives the base current of Q1 that charges the output load. The superiority of BiCMOS gates is based on the combination of the BJTs' high driving capability and the MOSFETs' very low static power consumption and high input impedance. The inverter in Figure 7.6 can easily be extended to perform complex multiple input logic functions in a similar way as for CMOS gates. For example, in the three inputs BiCMOS NOR gate M1 is replaced by three PMOS transistors in series and M2 by three NMOS transistors in parallel.

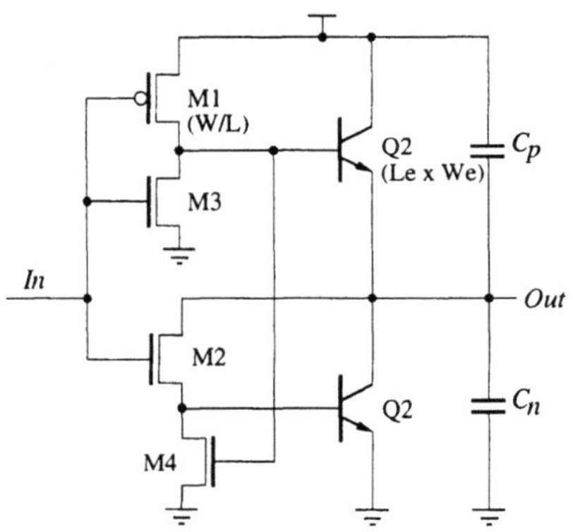

Figure 7.6

Conventional BiCMOS gate circuit.

The basic switching behavior of a conventional BiCMOS gate will now be analyzed. For a complete and accurate analysis we refer to [5] and [6]. The low to high output transition (Q1 charges the output load) is called pull-up transition, and the high to low transition (Q2 discharges the output load) is called pull-down transition. The logic levels for the inverter in Figure 7.6 are V_{DD}-V_{BEON} and V_{BEON}, respectively, where V_{BEON} is the BJT emitter to base ON threshold voltage. When the BJTs enter off state, there is a voltage drop between the output and V_{DD} or *Gnd* (depending on the output logic value) due to this ON threshold voltage. Taking the pull-up transition as a reference we will analyze the different phases of the switching transient. First, when there is an input transition (from high to low) a time elapses before M1 enters on state (the time taken by the input to go from V_{DD} to $V_{DD}-V_{TP}$). At this time M1 starts to charge the capacitances associated with the base of Q1. Then, a certain time later, the BJT switches on. At this time the current flowing through M1 is the gate current of Q1 and Q1 starts to charge the output capacitor by its emitter current. The output capacitance is made up of the load and the parasitic capacitance associated with the output that is mainly due to the capacitance between the collector of Q2 and the substrate. The main current that flows across the power supply terminals is the Q1 emitter current. The BJTs have two different behavior patterns depending on whether they are in high or low injection working regions. If we assume that Q1 is initially in low injection we find the following expression for the collector current:

$$I_c(t) = C_L^* \frac{dV_o(t)}{dt} = C_L^* \left(V_{DD} - V_{BEON}\right)\left[w_b + \frac{1}{w_b \tau_b^2}\right] \sin\left(w_b t\right) e^{-\frac{t}{\tau_b}} \quad \textit{Eq. (7.3)}$$

where $C_L^* = C_L + C_s$ is the effective output capacitance including the parasitic output node capacitance (C_s). Parameters w_b and τ_b are functions of the parameters of the BJT and the PMOS M1 and can be found in [5][6].

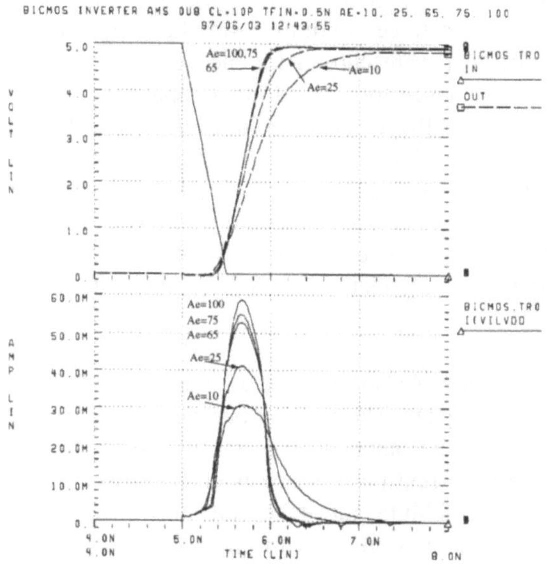

Figure 7.7

Simulation waveforms of a conventional BiCMOS inverter designed by using a 0.8μm BiCMOS technology, The figure shows the output voltage and supply current for different emitter areas of the BJTs. The optimum value is Ae = 65.

The current waveform of (7.3) can be approximated by a sinusoidal waveform if the exponential term can be reduced to unity (τ_b >> output switching time is a normal situation). Figure 7.7 shows the simulation waveforms of an output transition for a BiCMOS gate implemented with a 0.8μm technology loaded with 10pF. The gate is optimized to work in the limit of the high injection region. This corresponds to a Q1 and Q2 emitter area of 65 relative to the unity BJT device used in the technology. In the figure different gates have been plotted with the Q1 and Q2 emitter area as a parameter and high injection effects can be observed when they are smaller than 65. With bigger BJTs the performance of the gate is not improved.

We also analyzed the influence of input rise or fall time on current pulse shape. For lower rise and fall input times the slope of the rising edge of the current pulse is slightly increased. The effects of the input slope have only to be considered if the input switches very slowly [7].

7.1.3 Comparison of the switching noise generated by CMOS and BiCMOS gates

We will now compare the switching noise generated by two equivalent inverters, one of them CMOS and the other BiCMOS. To make a fair comparison the input capacitance and the delay for the same output capacitance may be equal for both gates [3]. However, asserting the three conditions simultaneously is not possible for the BiCMOS technologies that were chosen for the comparison. Figure 7.8 shows the delay versus output capacitance for a fixed input capacitance for the 1.2 μm and 0.8 μm technologies from AMS [8]. For each point of the curve the optimum size for the MOS transistors of a chain of 1, 2, 3 and 4 CMOS inverters and for the BJTs of a conventional BiCMOS inverter was calculated [9][10]. As shown in Figure 7.8, none of the curves for CMOS gates crosses the curve of the BiCMOS inverter. This means that the delay of the BiCMOS gates is always smaller than the delay of any CMOS tapered buffer (for these particular technologies, and without considering routing parasitics).

Though not shown here, it was also seen that an increase in the output load makes the current pulse wider and higher. The rising slope nevertheless remains the same (the shape is not changed). If the output load is raised too far the gate enters the high injection region and the current pulse shape is changed (see Figure 7.7). As already stated, a well-designed BiCMOS gate takes the maximum output capacitance into account and the emitter area is adjusted to attain optimum behavior with the BJTs working in the limit between high and low injection regions.

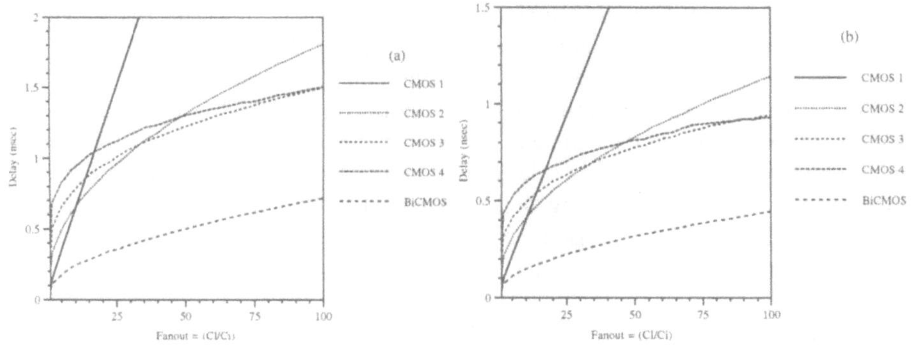

Figure 7.8

Optimum delay against output load for 1,2,3 and 4 stage tapered CMOS buffers and for a BiCMOS inverter with the same input capacitance. (a) 1.2 μm technology. (b) 0.8 μm technology

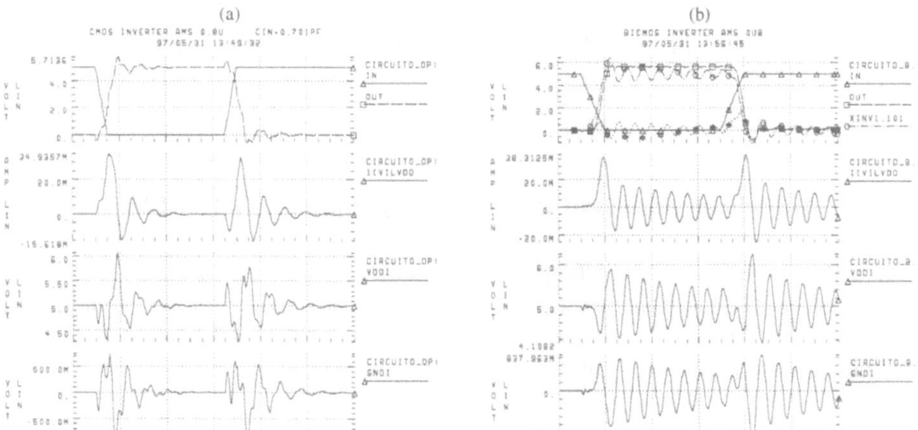

Figure 7.9

Voltage and current waveforms for the switching of (a) a CMOS inverter, and (b) a BiCMOS inverter.

The impossibility of satisfying the above three conditions meant that the comparison made was of an optimized CMOS inverter and a BiCMOS inverter with equal load capacitance and delay, but with different input capacitance. The input capacitance of the CMOS and the BiCMOS inverters was 701 fF and 191 fF, respectively. The load capacitance was 3.82 pF. This represents a fanout for the CMOS and the BiCMOS inverters of 3.67 and 20, respectively. The optimum emitter area for the BJTs of the BiCMOS inverters was found to be 48. The widths of the CMOS inverter NMOS and PMOS transistors were 102.7 μm and 289.84 μm, respectively. All the lengths are the minimum allowed by the design rules of the 0.8 μm technology used. Figure 7.9.a and Figure 7.9.b show the switching noise generated by the CMOS and the BiCMOS inverters, respectively. Input rise time is 0.5 ns. The circuit model of the package and the external power supply is the same one as in Figure 7.2.

The noise generated in both gates (CMOS and BiCMOS) is very similar, as can be observed in the simulations in Figure 7.9. The current pulse waveforms are not exactly the same. However, if it is assumed that half the charge used to switch the output node has already been transferred by the time the output reaches half the supply voltage ($V_{DD}/2$), and that the time taken to transfer this charge is the same for the two gates (they have the same delay) the maximum value of the current pulse and the half width of the current pulse must be almost the same in both gates. This means that the $dI(t)/dt$ is very similar for the two gates, as indicated by the simulation results. To conclude, it can be affirmed that from the viewpoint of switching noise generation neither of the two gate types (CMOS or BiCMOS) is any noisier than the other, with the same delay and output load.

7.2 Analytical models of the switching noise for CMOS ICs

The different models of switching noise presented in the relevant literature will be discussed in this section. Most of these models are based on the simultaneous switching of CMOS output drivers. However the models can also be applied to the switching noise generated by the internal CMOS circuitry.

7.2.1. Simultaneous switching noise (SSN) linear models

The first models for switching noise were published in the area of high speed ECL circuits and concentrated on the study of output drivers [17][21]. The switching noise spikes are modeled with a triangular pulse. Noise propagation and the influence of varying technology parameters are studied. In all these works the SSN is considered to be proportional to the number N of simultaneously switching drivers. The noise generated when the output drivers do not switch at exactly the same time is also studied. For example, if two drivers produce two noise pulses delayed a quarter of the base width of the triangular noise pulse used to model the SSN, the resulting switching noise pulse was only 60% of the amplitude that it would have been if the switching were exactly simultaneous. Noise pulse amplitude is reduced by 20% if the time separation between the switching of the two drivers is half the single pulse width.

7.2.2. Sub-linear SSN models and inclusion of short channel effects

The noise present in the power supply nodes modifies the voltage reference for the input and output voltage of the logic gates. The same noise thus affects itself with negative feedback. This feedback mechanism was first studied in [11]. Considering that the current waveform is a triangular pulse with half width equal to the rise or fall time of the input, the following expression for the SSN can be found:

$$V_n = V_k + \frac{t_r}{L_{eff} N\beta}\left[1 - \sqrt{1 + 2V_k \frac{L_{eff} N\beta}{t_r}}\right] \qquad \text{Eq. (7.4)}$$

where $V_k=V_{in}-V_t$ (the input voltage minus the threshold voltage), t_r is the time the current takes to rise from zero to its maximum value, L_{eff} is the inductance of the current loop from the power supply to the node that is charged or discharged by the output switching, $\beta = \mu_n C_{ox}(W / L)$ the gain of the NMOS or the PMOS transistor, depending on the output transition, and N is the number of simultaneously switching drivers. This expression has no dependence on the output load capacitance because it is assumed that its value is greater than the critical capacitance of (7.1).

The same authors later presented a model in which the current pulse waveform has a more realistic parabolic shape. This waveform is obtained if it is assumed that the noise increases linearly with the input slope. The expression for the SSN under these assumptions is:

$$V_n = V_k + \frac{V_k t_r}{2V_{DD} L_{eff} N\beta}\left[1 - \sqrt{1 + 4V_{DD} \frac{L_{eff} N\beta}{t_r}}\right] \qquad \text{Eq. (7.5)}$$

Here $V_k=V_{DD}-V_t$. The two previous models include the negative feedback effect that limits the noise increase with the number of simultaneous switching drivers. Another model of this type of higher complexity can be found in [12].

The above models use the Shichman-Hodges model for the MOS transistors. However in current submicron devices that are characterized by their low channel sizes short channel effects must be taken into account. The short channel effect that has most influence on the switching noise is the velocity saturation of the carriers. Long channel devices should work in saturation state or in ohmic zones during switching. Short channel devices designed to have fast transitions always work in saturation. In saturated short channel MOS transistors the current is not proportional to the square of the voltage, as it is in long channel devices. The current depends on the input voltage with an exponent smaller than 2. In heavily velocity-saturated devices this exponent is 1 and the current is directly proportional to the input voltage. If a realistic dependence between the saturation current and the input voltage is not considered, the simultaneous switching noise will be overestimated. The model in [13] includes the short channel effects and uses a variable exponent for the voltage dependence of the current called α_n [14]. The SSN expression obtained is:

$$V_n = s_r N\beta L_{eff} f\left[1 - e^{-\frac{(t-t_n)}{N\beta L_{eff} f}}\right] \quad t_n \le t \le t_r \qquad Eq.\ (7.6)$$

where s_r is the slope of the input waveform and t_n is the time taken by the NMOS or the PMOS (depending on the output transition direction) to enter ($t_n = s_r/V_T$) state. Parameter f must be empirically found and includes the dependence on exponent α_n. Parameter f is a constant if α_n is near unity. Product βf has admittance units [A/V]. Another model recently appearing using the same approach but including the effect of switching noise in the output of the non-switching gates is [15].

For heavily velocity-saturated devices, the above expression can be replaced by [16]:

$$V_n = \frac{s_r N\beta L_{eff} V_C}{2}\left[1 - e^{-2\frac{(t-t_n)}{N\beta L_{eff} V_C}}\right] \quad t_n \le t \le t_r \qquad Eq.\ (7.7)$$

where parameter f has been substituted by V_C, the critical lateral voltage above which velocity saturation occurs ($V_C = L_{eff} v_{sat}/\mu_{eff}$, the channel effective length, multiplied by the saturation velocity and divided by the effective mobility).

7.2.3 CMOS circuit parameters and switching noise

The parameters that influence switching noise can easily be identified from the model expressions given in the last subsection.

The dependence on supply voltage V_{DD} is present in all these models. Explicit dependence is always linear. However, the supply voltage is also implicitly present through the voltage dependence of the effective mobility of the β and the V_C. On one hand a drop in the supply voltage directly lowers the noise pulse amplitude but on the other hand increases the mobility that can to some extent compensate for the previous effect.

The transistors' dimensions determine the β value and if these are made smaller and smaller short channel effects will appear. In paper [13] the models presented in the previous subsection are compared with a SPICE simulation. For the 1.8 μm and 2.0 μm technologies the models described in (7.7) and (7.8) are accurate enough, but for 0.8 μm and 0.6 μm technologies the model used in (7.9) must be used to reach the same accuracy.

The input slope and the input rise or fall times directly affect the SSN. Reducing the input transition time increases the switching noise by the same amount. These transition times are not arbitrary and must be considered to be another technology parameter. They depend on the design of the logic gates and the input stages of the output drivers. In VLSI circuits the goal is to maximize the speed so the rising and falling times are the shortest possible in any case.

The last parameter considered is the number of simultaneous switching drivers or gates (N). There is a correction term that is subtracted from the lineal noise expression and that depends on N in both long and short channel models. Noise amplitude is affected sublinearly by N. This effect is of great importance if the generated switching noise level is high enough.

7.3 Switching noise transmission mechanisms

Switching noise generated by digital circuitry is present in the power supply nodes. This noise can be transmitted to other digital or analog circuits that are connected to the same power supply network. It is important to analyze the relationship between the design of the power supply distribution network and the transmission of switching noise, first of all. The switching noise generated in one packaged IC can also affect other ICs that communicate with it through the emitter and the receiver I/O drivers. A quiet (not switching) output driver of an IC supplied with a noisy power supply has a noisy output signal. The receiver in the other IC sees a noisy input signal that could cause functional errors and the input noise could even be propagated to the receiver's internal power supply nodes.

7.3.1 Power supply distribution in ICs

A multi-stage RLC network can be used to model the whole power supply network. The resistance and the inductance are due to the metal traces and vias and the capacitance is due to the decoupling capacitors inserted at the different levels (main power supply, board, package, and so on). Figure 7.10 shows a general diagram of the power supply distribution model, including the parasitic R and L of the decoupling capacitors. The design of the whole power distribution network will be discussed in a later chapter. We are now concerned only with the power distribution inside the IC. This on-chip power distribution network is mainly resistive.

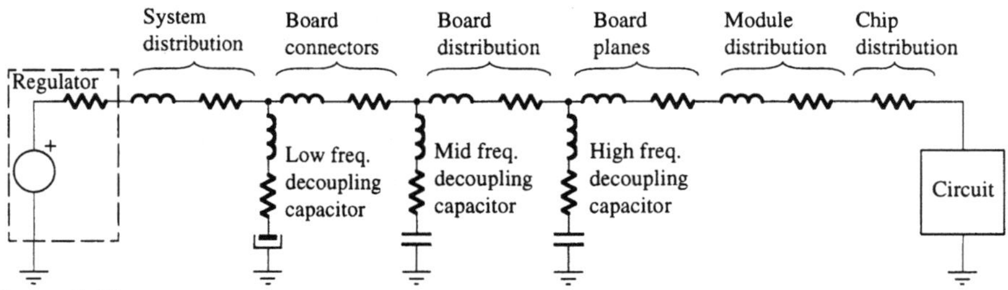

Figure 7.10

Power supply distribution structure for an electronic system. The decoupling capacitors include the parasitic inductance (ESL) and resistance (ESR) [17].

In [18] the different alternatives for distributing the power supply between several integrated blocks are analyzed. Inside the same IC there are noisy parts (mainly digital parts) and quiet parts, that may be either digital or analog blocks. In a first approach all the blocks are supplied with the same power bus. The structure of this power bus can be a tree, a star or a grid, as illustrated in Figure 7.11. The second approach is to supply the noisy and quiet blocks with different power buses (see Figure 7.12). Each power domain has its own distribution network that may be one of the three previous types. This last approach presents a theoretical isolation between the noisy and the quiet power domains at the expense of a bigger area and power supply pin count. Table 7.I summarizes the advantages and disadvantages of each of the power distribution alternatives. The symbols '↑' and '↓' are respectively used to symbolize an advantage or disadvantage concerning such aspects as the isolation or the routing. The symbols '+' and '-' are used to quantify the impedance or the area.

The best choice for minimizing switching noise is one that offers highest isolation. The tree structure produces high noise coupling due to its larger resistance. The grid and star structure have smaller power bus resistance and reduce noise coupling. The impedance of the power bus is important for three reasons. First, as said before, because a high impedance produces poor isolation. The other two reasons are connected with other techniques that can reduce switching noise. A low power bus resistance is an advantage for inserting on-chip decoupling capacitors. On the other hand, a high power bus resistance is beneficial by damping the oscillations produced by the RLC structure of the power supply distribution, including the package parasitic elements. The power bus resistance introduces a design trade-off between isolation and damping.

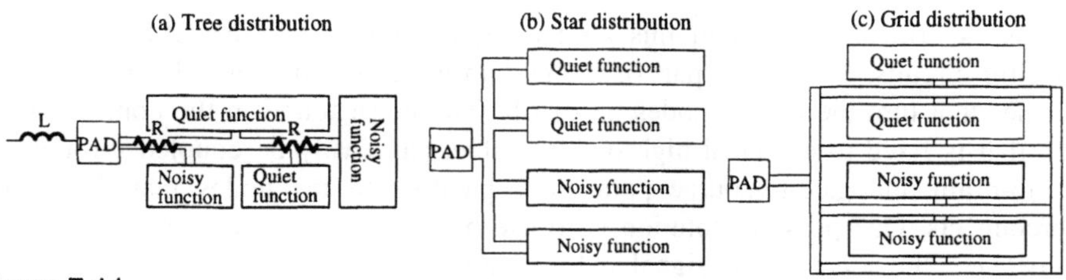

Figure 7.11

Power supply distribution alternatives [18].

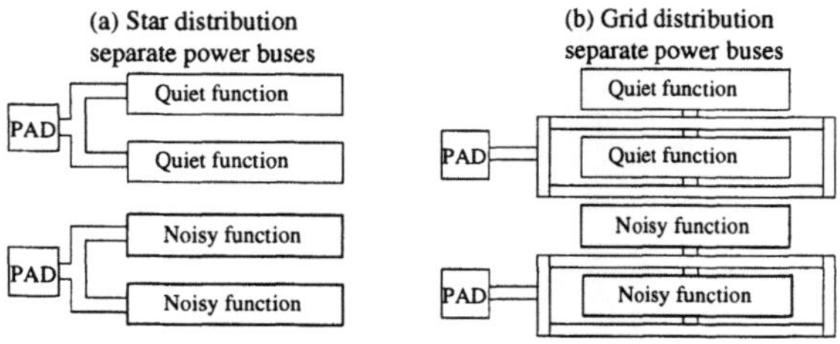

Figure 7.12

Power supply distribution alternatives with separate power supply buses for noisy and quiet functions [18].

In a circuit with separate power supply buses (often digital and analog buses) noise can be transmitted from one bus to the other. A block supplied, for example, by the noisy bus communicates with the other block supplied by the quiet bus. The outputs of the block supplied with the noisy power bus are also contaminated with the power supply noise. These noisy outputs can produce two different problems in the receiver: time jitter [17][19] and noise coupling to the quiet power bus. One example of the first problem is when a buffer in a noisy power supply domain receives a clock generated with a PLL in a quiet power bus domain. The clock distributed into the noisy domain has time varying rise and fall times because the switching threshold of the buffer changes with time due to the noise of its power supply. A solution for this problem involves using differential signaling between the noisy and the quiet domain. This is difficult in CMOS static logic circuits that are single ended. The second problem, coupling of the noise between the two separated power buses, is due to the parasitic capacitors between the input and the power supply nodes and between the input and the output nodes of the gates (in a CMOS receiver these are due to the MOSFETs' overlapping between the gate and the source and drain). To reduce this coupling the parasitic capacitance must be lowered. However, if several signals go from a noisy power domain to a quiet power domain the coupling of the noise is an additive process. A critical example of this kind of noise coupling is the CMOS switched capacitor circuits. The digital part of the system generates the control signals for the switches that turn the analog nodes on or off. The noise found in the control signal, even if it is fixed to a logical high or low value, can easily be coupled through the gate-to-drain and gate-to-source parasitic capacitors to the nodes that the switch interconnects. One possible solution is to use buffers supplied with the quiet power bus between the noisy control signal and the gate of the MOS switch [18].

Table 7.I.

Comparison between the different power supply distribution alternatives. The shaded cells indicate the best approach depending on each parameter.

	Single distribution			Split distribution	
Parameter	Tree	Star	Grid	Star	Grid
Area	-	+	++	++	++
Impedance	++	+	-	+	-
Routing	↓	↓	↑↑	↓	↑
Isolation	↓↓	↑	↓	↑↑	↑↑

7.3.2 Substrate biasing

Substrate noise coupling has been analyzed at length in the previous chapters. There are two ways noise is injected into the substrate. The first one is capacitive coupling across the PN junction of the devices and the substrate (or well). The second one is direct injection of the switching noise found in the power supply nodes to the substrate across the biasing contacts. The most important of these proved to be the second one [20]. Noise is received in three different ways: the previous two and the body effect of the MOSFETs that produce variations in the threshold voltage due to the noise-induced time-varying substrate voltage.

Solutions for this kind of switching noise coupling will be found in the relevant chapters of this book.

7.3.3 Communication between two ICs

Switching noise can be propagated to other parts of the system, even outside the IC where the noise is generated. This is the case of the chip to chip signals from an IC with noisy power supplies and a quiet IC [21][22]. The output drivers of an IC are usually made of a chain of progressively sized inverters, and designed to switch high output loads very quickly. This means very high and very short switching current pulses. All the outputs of a bus could switch simultaneously producing a huge amount of *dI/dt* noise in the power lines of the output drivers. It is therefore usual to have separated power lines, pads and pins for the output drivers and for the rest of the IC. This reduces the amount of output driver switching noise inside the IC but does not reduce the noise that could be coupled to other ICs through the outputs. Switching noise makes the supply voltage fluctuate with time. This in turn makes the falling and rising edges of the outputs time-varying, producing delay errors in the receivers. Switching noise is also coupled to the output of the non-switching drivers. These undesired voltage pulses are propagated to the receiver inputs and may cause logic errors or be coupled to the receiver's supply lines. Usually the supply lines of input receivers are the same as for the rest of the circuit, so the noise is coupled to the whole receiver IC circuitry through the input driver. The best solution for this problem is to use differential chip to chip signaling [23], but this doubles the width of the buses.

7.4 The effects of switching noise in digital circuits

The first studies on the effects of switching noise concentrated on purely digital circuits [17][21][24]. In digital circuits switching noise can produce delay error in the output nodes that are switching and false switching in the nodes that are quiet. Wrong values can be stored in the latches and flip-flops, producing permanent errors. These effects are more important in dynamic logics, like for example domino logic [25]. In the dynamic gates there are floating nodes that can easily be discharged due to the noise of the power supply [26]. In this section all these effects in static CMOS logic will be analyzed.

7.4.1 Delay errors

Modification of normal transition times is due to the change in the current caused by reducing or increasing the gate-to-source voltage of the output transistors due to the noise in its source terminals (the voltage supply nodes). A simulation experiment was performed to analyze the delay changes produced by switching noise. Figure 7.13 shows the circuit used in the experiment. There are two inputs, one for the noise generators (*ia*), which are chains of simultaneously switching inverters, and the other for the test inverter (*ib*). The two inputs can change in the same or the opposite direction. Transition of the generators can be arbitrarily delayed to have a voltage spike at the same time as the test inverter switches. The noise spikes always have an opposite sign in the two internal supply nodes (*vddi* and *gndi*). Figure 7.14 shows the simulated waveforms for two cases. In Figure 7.14.a the *ia* and *ib* transitions are in counter phase and a negative voltage spike in vddi and a positive one in gndi occur at the same time as the output transition of the test inverter. The transition delay is greater than the normal delay without switching noise. In Figure 7.14.b *ia* and *ib* are also in counter phase but they are delayed in such a way that now the voltage spikes are positive at the vddi node and negative at the gndi node when the output of the test inverter switches. In the last case the delay is reduced in respect of the normal value.

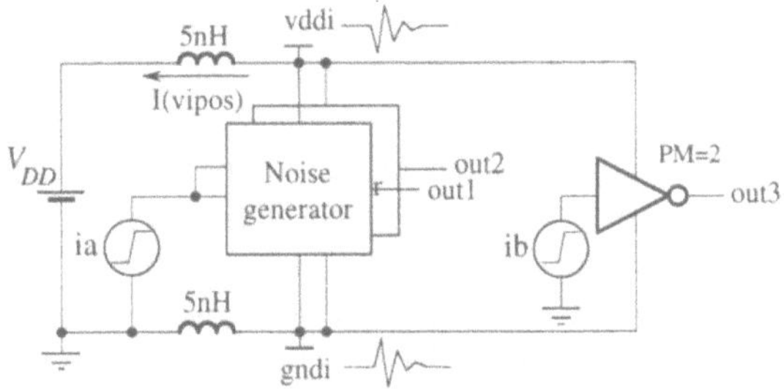

Figure 7.13

Circuit used to simulate the effects of switching noise in the output delay of an inverter.

Table 7.II summarizes the results of the experiment. For each of the four cases (*ia* and *ib* in phase or counter phase and different noise spike signs) the low to high (tp_{LH}) and high to low (tp_{HL}) propagation times are measured in the simulation. The noise spike amplitude is not too high (500 mV), but the mean delay variation is about 0.4 ns and the maximum variation is 1 ns.

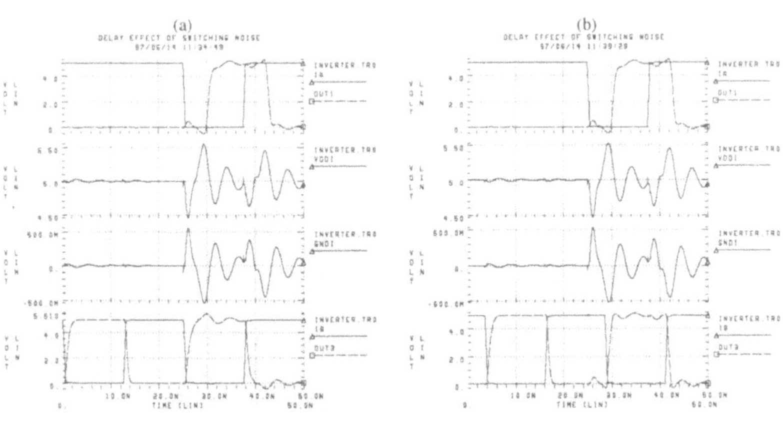

Figure 7.14

Simulated waveforms of the effects of switching noise on the delay of a CMOS inverter. (a) Increase in the delay due to a positive noise pulse in gndi and a negative noise pulse in vddi. (b) Decrease in the delay due to a negative noise pulse in gndi and a positive noise pulse in vddi.

Table 7.II.

Changes in the delay of a CMOS inverter due to switching noise obtained by simulation. The symbols '+' and '-' are used to indicate the polarity of the noise pulse in the internal supply nodes. All time values are in ns.

Ia/ib	noise *gndi*	noise *vddi*	Tp_{LH} Normal	tp_{LH} with noise	Δtp_{LH}	tp_{HL} normal	tp_{HL} with noise	Δtp_{HL}
phase	+	-	6.7	7.18	0.48	5.24	5.64	0.40
phase	-	+	6.7	6.31	-0.39	5.24	4.66	-0.42
Counter-phase	+	-	6.7	6.92	0.22	5.24	6.25	1.10
Counter-phase	-	+	6.7	6.37	-0.27	5.24	4.60	-0.64

7.4.2 False switching

The noise disturbance from power supply nodes can easily be transmitted to the outputs of gates that are not switching. This noise at the output nodes of the gates can be transmitted to other gates and make them switch if the noise levels are greater than the switch threshold voltage of the gates. There are two successive processes involved: transmission of the noise from the supply nodes to the output of a gate and later transmission of this noise to other gates. The first process is easy to explain for static CMOS gates. For example, if the output is at '0' logic level, at least one of the NMOS transistors of the gate connects the output node to the noisy internal gnd supply node. The NMOS transistor is working in the ohmic region and any disturbance at the source is transmitted to the drain, the output of the gate [27]. The second process, the transmission of this noise through a logic circuit, has been the subject of several pieces of work, connected with the dynamic noise margins of logic gates [28]. A logic gate does not propagate the noise from its input to its output if the input noise levels are smaller than the noise margins of the gate. The noise margin for '0' is usually different from the noise margin for the '1'.

These noise margins depend on the width of the input noise pulse. A very short input noise pulse could have a higher amplitude than a wider one without being propagated by the gate. For a fixed noise pulse amplitude there is a certain pulse width above which the pulse is propagated. These dynamic noise margins for ECL logics are studied in [21]. The same analysis is made for dynamic CMOS logic gates in [26]. In both cases the noise margins are small and the switching noise coupled to the outputs can easily produce false switching in other gates. In [29] the same analysis for CMOS receivers with TTL compatible logic levels is put forward. In this case the noise margin for the '0' logic level is small because the switching threshold for TTL logic circuits is in the 1.2V–1.6V range. Apart from this, conventional static CMOS logic gates have high immunity against noise pulse propagation, as was studied in [30][31].

7.4.3 Erroneous storage into flip-flop and latches

One of the worst effects of switching noise in digital circuits is the erroneous storage of information in flip-flops [32]. The two previous noise effects affect latches and flip-flops. Delay variations could be the cause of set-up and hold time violations[2] [33]. False switching of the clock or data inputs of the flip-flop could cause erroneous storage. Even noise from the supply nodes could cause these errors. In dynamic logics the problem is particularly serious because a noise pulse in the power supply terminal of the gate with an amplitude equal to the threshold voltage of the MOSFETs could discharge the dynamic nodes [26][34].

7.4.4 Sensitivity of BiCMOS logic gates to switching noise

BiCMOS logic gates are less sensitive to switching noise from power supply nodes than CMOS logic gates. The base-to-emitter and base-to-collector junctions of pull-up and pull-down bipolar transistors, when these are inverse biased, block out noise transmission from the power supply nodes to the output of the gate. BJTs are furthermore unidirectional devices, which reduces the number of cases in which noise can be propagated through them. The noise coupled from the power supply node to a quiet BiCMOS gate depends on the logic value of its output and also on the power supply terminal where the noise is found, because the gate structure is not symmetrical. The worst case is when the output is at low logic level and there is a positive noise pulse at internal Gnd node. As the pull-down transistor is ON the output follows the Gnd node with a direct biased diode voltage drop [35]. BiCMOS gates are thus less sensitive to switching noise and have switching noise-induced output noise pulses of less amplitude. This reduces false switching and other problems derived from this.

[2] These two time constraints are due to the finite nature of the slope of the clock signal. The set-up time is defined as the minimum time between a change in the input and the clock edge of the flip-flop. The hold time is defined as the minim time that the input has to remain stable after the edge of the flip-flop. A violation of any of these two time constraints could cause an erroneous storage in the flip-flop.

7.5 The effects of switching noise in analog circuits

Noise is one of the most important limitations to the performance of analog systems. dI/dt switching noise is the main source of noise in current ICs and, as will be seen further on, the problem will become worse in future ICs. In mixed signal ICs the noise generated by the digital part easily affects the analog part. There are several elements common to both kinds of circuits that are paths for transmitting switching noise. The substrate, the power supply lines and the signals that communicate digital and analog cells are the most important of these. Noise transmission across the two last elements could be avoided by using separate analog and digital power supply buses and differential signals between analog and digital circuits. However, substrate transmission will still remain. The greatest efforts to reduce noise problems in mixed signal ICs have been made in the field of substrate noise propagation [36][37][38].

The main effects of switching noise on analog circuits are connected with the degradation of the signal to noise ratio (SNR). In [39], for example, the design process of a communications system in which the level of the input signal is about 70 nV is presented. The mean noise value found in the internal power supply nodes reaches 40 mV (the paper gives no information about the final magnitude of noise at the input of analog circuits). These figures can help us understand the magnitude of the problem. SNR reduction in mixed signal ICs requires better design techniques and more complex analog circuits, increasing the cost and power consumption, two of the most important design constraints in mobile communication systems. Other examples of these problems can be found in [40][41].

Two important analog basic blocks are analog-to-digital converters (ADCs) and operational amplifiers (OPAMPs). Both may be affected by switching noise. In [42] some measurements of the effects of switching noise in OPAMPs are given. The noise produces an increase in distortion in the output waveform as well as phase errors. In ADCs the most important effect is the limitation in the minimum signal level that can be distinguished from the noise. This limit fixes the maximum converter resolution [43][44][45][46]. In short, the switching noise that is transmitted to the analog part of the mixed signal IC makes it hard to work with low input signal levels, limiting the dynamic range of analog circuits. The noise leads to distortion in the signals amplified by OPAMPs and causes phase errors that could severely affect linear filters or demodulators. The failure induced by switching noise in the analog part of mixed signal ICs is not functional, as it was in digital circuits, but involves a reduction and limitation of the performance of analog circuits.

7.6 Conclusions

This chapter compiles the state of the art of the switching noise phenomenon. Our analysis focused on on-chip switching noise, produced inside the packaged IC. In this chapter the package was modeled by the effective inductance of its power supply pins.

Static CMOS and BiCMOS gates were analyzed and compared. There are no significant differences in the magnitude of the noise generated by both types of gates. The different factors that modify the amount of noise generated by a gate were also analyzed. On logic level these are the input combination, the fanout of the gate and the number of gates that switch simultaneously. On transistor level these are the supply voltage, the size of the transistors, the slope of the input waveforms and the output load of the gate. Several analytical models of on-chip logic switching noise were presented, some of these including short channel effects. The noise generated by an increasing number of simultaneous switching gates is not proportional to the number of gates, but increases sublinearly.

Power supply buses, common substrate and signals that communicate the noisy and quiet parts of the IC, like the digital and the analog parts, transmit switching noise. The most important of these transmission mechanisms is the substrate, because the other transmission channels can be minimized (by using separate power supply buses for noisy and quiet parts and differential signaling between noisy and quiet blocks).

Digital static CMOS logic is very robust against switching noise. Nevertheless, other kinds of digital logics (such as ECL logics, TTL compatible logics and moreover dynamic CMOS logics) are more sensitive to this. The effects that switching noise could produce in digital circuits are: logical errors (transient errors such as false switching or delay errors) and permanent errors such as wrong storage in flip-flops and latches. This is becoming an important research topic in the area of testing and verifying logic circuits because noise-induced faults must be detected to assure good reliability [47][48]. In the analog parts of the IC switching noise produces a limitation in performance by reducing the dynamic range of OPAMPs or introducing phase errors and degrading the SNR at the input stage of the ADCs, which limits the maximum resolution possible.

References

[1] Arun Vaidyanath, Birgir Thoroddsen, and J.L. Prince, 'Effect of CMOS Driver Loading Conditions on Simultaneous Switching Noise', *IEEE Tr. on Components, Packaging, and Manufacturing Technology-Part B*, Vol. 17, No. 4, pp. 450–485, 1994.

[2] Yi-Shing Chang, Sandeep K. Gupta, and Melvin A. Breuer, ' Analysis of Ground Bounce in Deep Sub-Micron Circuits', private communication.

[3] Prasad Raje, 'Digital Design' in *BiCMOS Technology and Applications*, A.R. Álvarez (ed), Kluwer Academic Publishers, Massachusetts, 1993.

[4] M.I. Elmasry (ed), *BiCMOS Integrated Circuit Design*, IEEE Press, Piscataway, 1994.

[5] P.Raje, K. Saraswat, and K. Cham, 'Accurate Delay Models of Digital BiCMOS', *IEEE Trans. Electron Devices*, Vol. 39, No. 6, 1992.

[6] Vicente Jiménez, and Ramón Alcubilla, 'An Accurate Transient Model for BiCMOS Gate Optimization', *Int. J. Electronics*, Vol. 82, No. 5, pp. 503–514, 1997.

[7] Shayan Zhang, T.S. Kalkur, 'Analysis of BiCMOS Buffer for Input Voltages with Finite Rise Time', *IEEE J. of Solid-State Circuits*, Vol. 29, No. 7, pp. 797–807, 1994.

[8] 0.8 μm and 1.2 μm BiCMOS Process Parameters. Austria Mikro Systeme International. Schloß Premstätten, A-8141 Unterpremstätten, Austria.

[9] Wen Fang, Arthur Brunnschwiler, and Peter Ashburn, 'An Accurate Analytical BiCMOS Delay Expression and its Application to Optimizing High-Seed BiCMOS Circuits, *IEEE J. of Solid-State Circuits*, Vol. 27, No. 2, pp. 191–202, 1992.

[10] Prasad A. Raje, Krishna C. Saraswat, and Kit M. Cham, 'A New Method for Design of BiCMOS Gates and Comparison with CMOS', *IEEE Tran. Electron Devices*, Vol. 39, No. 2, pp. 339–347, 1992.

[11] R. Senthinathan, and J. L. Prince, 'Simultaneous Switching Ground Noise Calculation for Packaged CMOS Devices', *IEEE J. of Solid-State Circuits*, Vol. 26, No. 11, pp. 1724–1728, November 1991.

[12] Yaochao Yang, and John R. Brews, 'Design Trade-Offs for the Last Stage of an Unregulated, Long-Channel CMOS Off-Chip Driver with Simultaneous Switching Noise and Switching Time Considerations', *IEEE Tran. on Components, Packaging, and Manufacturing Technology—Part B*, Vol. 19, No. 3, pp. 481–486, August 1996.

[13] Srinivasa R. Vemuru, 'Accurate Simultaneous Switching Noise Estimation Including Velocity-Saturation Effects', *IEEE Transactions on Computers, Packaging, and Manufacturing Technology—Part B*, Vol. 19, No. 2, pp. 344–349, May 1996.

[14] Takayasu Sakurai, and A. Richard Newton, 'Alpha-Power Las MOSFET Model and its Applications to CMOS Inverter Delay and Other Formulas', *IEEE J. of Solid-State Circuits*, Vol. 25, No. 2, pp. 584–593, April 1990.

[15] Hye-Ran Cha and Oh-Kyong Kwon, 'A New Analytic Model of Simultaneous Switching Noise in CMOS Systems', *Electronic Components and Technology conference*, pp. 615–621, May 1998.

[16] Yaochao Yang, and John R. Brews, 'Design for Velocity Saturated, Short-Channel CMOS Drivers with Simultaneous Switching Noise and Switching Time Considerations', *IEEE J. of Solid-State Circuits*, Vol. 31, No. 9, pp. 1357–1360, September 1996.

[17] E.E. Davidson, 'Electrical Design of a High Speed Computer Package', *IBM J. of Research and Developement*, Vol. 26, No. 3, pp. 349–361, 1982.

[18] Nishath K. Verghese, Timothy J. Schmerbeck, David J. Allstot, 'Chapter 8: Modeling Chip/Package Power Distribution' in 'Simulation Techniques and Solutions for Mixed-Signal Coupling in Integrated Circuits', Kluwer Academic Publishers, Boston, 1995, pp. 149–182.

[19] Perng-Shyong Lin, and A. Zukowski, 'Analysis and Control of Timing Jitter in Digital Logic Arising from Noise Voltage Sources', *IEEE Int. Conf. Computer Design*, pp. 352–356, 1993.

[20] Xavier Aragones, 'Caracterització de l'acoblament per substrat' en *Contribució a l'Estudi de l'Acoblament per Substrat en Circuits Integrats Mixtes*, Tesis Doctoral, Dept. d'Enginyeria Electrònica, Univ. Politècnica de Catalunya, 1997.

[21] George A. Katopis, 'Delta-I Noise Specification for a High-Performance Computing Machine', *Proc. of the IEEE*, Vol. 73, No. 9, pp. 1405–1415, September 1985.

[22] Michael Dolle, 'Analysis of Simultaneous Switching Noise', *IEEE Symp. Circuits and Systems*, pp. 904–907, 1995.

[23] A. J. Rainal, 'Eliminating Inductive Noise of External Chip Interconnections', *IEEE J. Solid-State Circuits*, Vol. 29, No. 2, February 1994, pp. 126–129.

[24] E.E. Davidson, P.W. Hardin, G.A. Katopis, M.G. Nealon, L.L. Wu, 'Physical and Electrical Design Features of the IBM Enterprise System/9000 Circuit Module', *IBM J. of Research and Development*, Vol. 36, No. 5, pp. 877–888, September 1992.

[25] R.H. Krambeck, Charles M. Lee, and Huang-Fai Stephen Law, 'High-Speed Compact Circuits with CMOS', *IEEE J. of Solid-State Circuits*Vol. SC-17, No. 3, pp. 614–619, June 1982.

[26] Patrik Larsson, and Christer Svensson, ' Noise in Digital Dynamic CMOS Circuits', *IEEE J. of Solid–State Circuits*, Vol. 29, No. 6, pp. 655–662, June 1994.

[27] Mary Sue Haydt, Robert Owens, and Samiha Mourad, 'Modeling the Effect of Ground Bounce on Noise Margin', *Porc. International Test Conference 1994*, pp. 279–285.

[28] Jacek M. Zurada, Yu Sang Joo, Samuel V. Bell, 'Dynamic Noise Margins of MOS Logic Gates', *Proc. of ISCAS' 89*, pp. 1153–1156.

[29] R. Senthinathan, J.L. Prince and S. Nimmagadda, 'Noise Immunity Characteristics of CMOS Receivers and Effects of Skewing/Damping CMOS Output Driver Switching Waveform on the Simultaneous Switching Noise', *Microelectronics Journal*, Vol. 23, 1992, p. 29–36.

[30] Francesc Moll, and Antonio Rubio, 'Analysis of Spurious Signal Propagation in Digital CMOS Circuits', *Int. J. Electronics*, Vol. 73, No. 3, pp. 621–625, 1992.

[31] Francesc Moll, and Antonio Rubio, 'Spurious Signals in Digital CMOS VLSI Circuits: A Propagation Analysis', *IEEE Tran. on Circuits and Systems–II*, Vol. 39, No. 10, pp. 749–752, October 1992.

[32] Francesc Moll Echeto, 'Parasitic Effects Due to Interconnections in Microelectronic Design', Dissertation for the Ph.D. degree, Electronic Engineering Department, Universitat Politècnica de Catalunya, 1995.

[33] S. H. Unger, and C.-J. Tan, 'Clocking Schemes for High-Speed Digital Systems', *IEEE Tran. on Computers*, Vol. C-35, No. 10, October 1986.

[34] Daniel W. Dobberpuhl, 'A 200-Mhz 64-b Dual-Issue CMOS Microprocessor', *IEEE J. of Solid-State Circuits*, Vol. 27, No. 11, pp. 1555–1564, November 1992.

[35] David A. Secker, and John L. Prince, 'Effects and Modeling of Simultaneous Switching Noise for BiCMOS Off-Chip Drivers', *IEEE Tran. on Components, Packaging, and Manufacturing Technology—Part B*, Vol. 19, No. 3, pp. 474–480, August 1996.

[36] Balsha R. Stanisic, Rob A. Rutenbar, and L. Richard Carley, 'Addressing Noise Decouling in Mixed-Signal IC's: Power Distribution Design and Cell Customization', *IEEE J. of Solid-State Circuits*, Vol. 30, No. 3, pp. 321–326, March 1995.

[37] Sujoy Mitra, Rob. A. Rutenbar, L. Richard Carley, and David J. Allstot, 'Substrate-Aware Mixed-Signal Macrocell Placement in WRIGHT', *IEEE J. of Solid-State Circuits*, Vol. 30, No. 3, March 1995.

[38] Martin Pfost, Hans-Martin Rein, and Thomas Holzwarth, 'Modeling Substrate Effects in the Design of High-Speed Si-Bipolar IC's', *IEEE J. of Solid-State Circuits*, Vol. 31, No. 10, pp. 1493–1501, October 1996.

[39] Patrik J. Zabinski, Mark E. Cickberg, Barry K. Gilbert, Philip J. Zucarelli, Darren V. Weninnger, Theodore W. Keller, and David M. Yee, 'Case Study: The Design of a Mixed-Signal Global Positioning System Receiver Using Multichip Module Packaging', *IEEE Tran. on Components, Packaging, and Manufacturing Technology—Part B*, Vol. 19, No. 1, pp. 215–224, February 1996.

[40] A BiCMOS Time Interval Digitizer Based on Fully-Differential, Current-Steering Circuits', *IEEE J. of Solid-State Circuits*, Vol. 29, No. 6, June 1994, pp. 707–714.

[41] K.J. Schultz, R.G. Gibbins, J.S. Fujimoto, R.S. Philips, G.F.R. Gibson, A.L. Silburt, 'Low-Supply-Noise Low-Power Embedded Modular SRAM', *IEE Proc..-Circuits, Devices and Systems*, Vol. 143, No. 2, April 1996, pp. 73–82.

[42] Johan Catrysse, 'Measured Distortion of the Output-Waveform of an Integrated OPAMP Due to Substrate Noise', *IEEE Tran. on Electromagnetic Compatibility*, Vol. 37, No. 2, pp. 310–312, May 1995.

[43] B.M.J. Kup, E.C. Dijksmans, P.J.A. Naus, J. Sneep, 'A Bit Stream Digital-to-Analog Converter with 18-b Resolution', *IEEE J. of Solid-State Circuits*, vol. 26, no. 12, pp. 1757-1763, December 1991

[44] Brian P. Brandt, and Bruce A. Wooley, 'A 50-MHz Multibit Sigma-Delta Modulator for 12-b 2-MHz a/D Conversion', *IEEE J. of Solid-State Circuits*, Vol. 26, No. 12, pp. 1746–1756, December 1991.

[45] J.W. Fattaruso, S. Kiriaki, M. de Wit, C. Warmar, 'Self-Calibrating Techniques for a Second-Order Multibit Sigma-Delta Modulator', *IEEE J. of Solid-State Circuits*, Vol. 28, No. 12, pp. 1216–1223, December 1993.

[46] Tallis Blalack, and Bruce A. Wooley, 'The Effects of Switching Noise on an Oversampling A/D Converter', *1995 IEEE International Solid-State Circuits Conference*, pp. 200–201, 367.

[47] Alan Rogers, 'Reliability Issues Due to Switching Currents in CMOS ASICs', *Proc. NORTHCON '89 Conference*, pp. 304–309.

[48] Melvin A. Breuer, and Sandeep K. Gupta, 'Process Aggravated Noise (PAN): New Validation and Test Problems', *Proc. 1996 International Test Conference*, pp. 914–923.

8 Digital design for low switching noise

This chapter looks at the relationship between the digital design process and the switching noise generated by the circuits finally obtained. This analysis is performed on all levels: abstract, structural, logic, and circuit. High-level synthesis and logic synthesis processes are examined and some low switching noise design rules are derived. Switching noise in output drivers is tackled by proper design techniques that are summarized at the end of the chapter

8.1 Introduction

The process of designing complex digital circuits starts with a functional (also called behavioral) description of the circuit. High-level description languages like VHDL [1] or Verilog [2] are used to formally represent the circuits on this functional level. Then CAD tools synthesize the circuit to get a logical or structural representation. The designer may introduce area, timing or power consumption as design constraints in this process. The high-level synthesis process usually involves two different options: attempting to maximize speed or to minimize area. These alternatives are analyzed in the first section of the chapter.

The structure of the synthesized circuit is also a design decision that may influence circuit activity and thus the switching noise generated. This question is dealt with in the second section, along with the logic synthesis process.

The logic circuit can be implemented in different ways using the logic alternatives available. These logic alternatives (static CMOS, dynamic logic, self-timed logic, etc.) are analyzed in section three.

The following section contains a study of the relationship between the low power design approaches existing in literature and the reduction of switching noise.

The chapter ends with a summary of the design techniques used to minimize switching noise in output drivers.

8.2 High level design and synthesis *versus* switching noise

Switching noise is closely related to circuit structure, but this structure is obtained to implement a given functionality. The structure needed to fulfill any particular function is not unique. There may be some structures that are more suitable for generating low levels of switching noise than others with the same functionality. Switching noise at these functional or structural levels is related to simultaneous switching activity, as the target variable to be minimized.

The design constraints introduced in the synthesis process can be used to obtain better structures from the switching noise point of view. In the following subsections the synthesis design process is analyzed and simple design rules are derived from this and illustrated with synthesis examples.

8.2.1 High level synthesis process

The starting point for the synthesis process is a file containing the high level description of the circuit using any of the high-level description languages (VHDL, Verilog, HardwareC, etc.). The first step is a syntactic analysis and a translation into an internal representation of the circuit, usually in the form of a data-flow graph (DFG). The second step in the synthesis process generates the structure of the circuit from the DFG, applying some constraints. The third step consists of the logic synthesis and the technology mapping of the structure into a library of logic cells.

Switching noise reduction can be tackled on different levels in the synthesis process. At the architecture synthesis stage scheduling and resource binding algorithms define the final structure of the circuit. These algorithms use criteria such as area minimization or delay minimization as constraints. The smaller area is obtained by a resource-limited synthesis, but this nevertheless increases the longest path delay. The minimum delay solution is obtained by proper scheduling that tends to maximize the concurrence of the structure. Between these two extremes, synthesis algorithms can be modified to introduce newer criteria. For example, the scheduling can be undertaken in such a way that, once the maximum time depth of the DFG is stabilized, the operations are distributed in an attempt to minimize the number of operations performed at a given time step. The control unit synthesis can also be performed using low noise design criteria, such as the minimization of the number of bits that change from one state to the next, by using Gray state encoding. Other strategies used to maximize the throughput are not indicated from the low switching noise standpoint, such as pipeline structures, as we will see further on. The logic synthesis process will also influence the final structure of the circuit and low noise logic synthesis constraints can be introduced. This will be dealt with in the next subsection.

With today's CAD synthesis tools the designer has to make a decision between the area and the timing constraints. The synthesis can be performed to get minimum area (also called minimum cost solution) or maximum speed (minimum delay solution). The following examples show how minimum cost solutions are better suited to low switching noise design than maximum speed solutions.

8.2.2 Example 1: Traffic light controller

The circuit used in this example is one of the VHDL benchmarks available from 1992 [3]. The circuit is a finite state machine and has been synthesized using the Synergy software from Cadence [4]. The high level description of the circuit is shown in Figure 8.1. Two final implementations were synthesized: one with minimum cost (TLCcost) and another with maximum speed (TLCtime) that are shown in Table 8.I and in Figure 8.2.

```
----------------------------------------------------------------------------
--
-- Traffic Light Controller (TLC)
--
-- Source:  Hardware C version  written by David Ku on June 8, 1988 at Stanford
--
-- VHDL Benchmark author Champaka Ramachandran
--                       University Of California, Irvine, CA 92717
--                       champaka@balboa.eng.uci.edu
--
-- Developed on Aug 11, 1992
--
Verification Information:

--                   Verified     By whom?             Date          Simulator
--                   --------   ------------         --------      ------------
--  Syntax             yes    Champaka Ramachandran  Aug 11, 92     ZYCAD
--  Functionality      yes    Champaka Ramachandran  Aug 11, 92     ZYCAD
------------------------------------------------------------

entity TLC is
        port (
              Cars : in  BIT;
              TimeoutL : in  BIT;
              TimeoutS : in  BIT;
              StartTimer : out  BIT;
              HiWay : out  BIT_VECTOR(1 downto 0);
              FarmL : out  BIT_VECTOR(1 downto 0);
              state : out  BIT_VECTOR(1 downto 0);
                clk: in bit
            );
end TLC;


architecture TLC of TLC is

  signal newstate, state_aux, current_state : BIT_VECTOR(1 downto 0);
  signal newHL, newFL : BIT_VECTOR(1 downto 0 );
  signal newST : BIT;

begin
------------------------------------------------------------
traffic:process(current_state, Cars, TimeoutS, TimeoutL)
```

```
begin
--   current_state := newstate;
-- combinational logic to determine nextstate
   case current_state is
     when "00" =>  newHL <= "10";  newFL <= "11";
                          if (Cars = '1') and (TimeoutL = '1') then
                          newstate <= "10"; newST <= '1';
                          else
                            newstate <= "00"; newST <= '0';
                          end if;
     when "10" =>  newHL <= "01"; newFL <= "11";
                           if (TimeoutS = '1') then
                             newstate <= "01"; newST <= '1';
                           else
                             newstate <= "11"; newST <= '0';
                           end if;
     when "01" =>    newHL <= "11"; newFL <= "10";
                           if (Cars = '0') or (TimeoutL = '1') then
                               newstate <="11"; newST <= '1';
                            else
                             newstate <= "01"; newST <= '0';
                           end if;
     when "11" =>   newHL <= "11"; newFL <= "01";
                           if (TimeoutS = '1') then
                             newstate <= "00"; newST <= '1';
                            else
                             newstate <= "11"; newST <= '0';
                           end if;
     when others =>
                           newstate <= "00";
                           newHL <= "00";
                           newFL <= "00";
                           newST <= '0';
     end case;
--  wait for 10 ns;
end process traffic ;
------------------------------------------------------------
current_state <= state_aux;
state <= state_aux;
fsm:process
        begin
                 wait until (clk'event and clk='1');
                 state_aux <= newstate;
                 HiWay <= newHL;
                 FarmL <= newFL;
                 StartTimer <= newST;
        end process;
end TLC;
```

Figure 8.1

VHDL description of the traffic light controller circuit (TLC).

Table 8.1.

Results of the two synthesised solutions of the TLC circuit.

Parameter	TLCcost	TLCtime
Number of cells	22	25
Area (μm^2)	8553.60	9375.68
Maximum frequency	223.52 MHz	311.95 MHz
Longest path delay	4.47 ns	3.21 ns

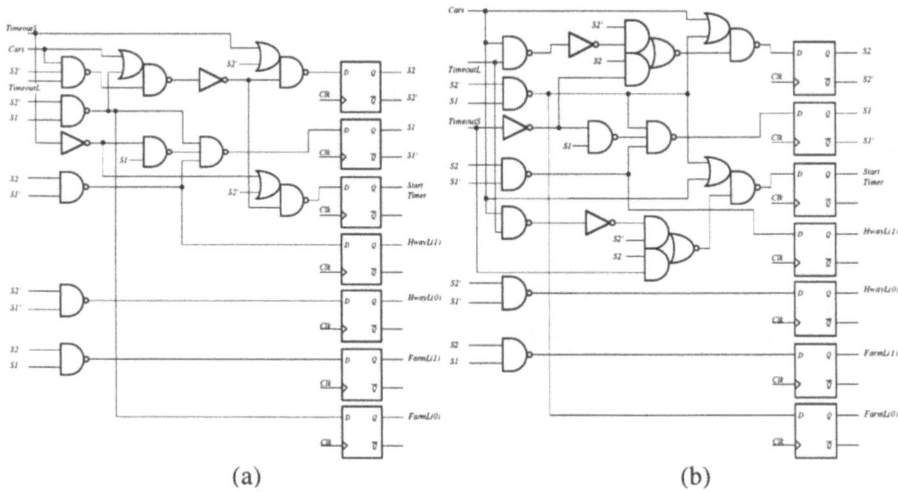

Figure 8.2

Circuit schematic of the two synthesised solutions: (a) TLCcost and (b) TLCtime. The gates are from the 0.8 μm AMS CMOS cell library.

The two synthesized solutions are only slightly different because the circuit specification of the TLC is quite simple. The next step consists of a Verilog [5] simulation where the simultaneous switching activity (number of transitions in every node of the circuit) is obtained for a given input pattern. The results are shown in Figure 8.3. The switching activity of the TLCcost circuit is slightly smaller than the switching activity of the TLCtime circuit, as expected. In order to validate the digital simulation process used to obtain the simultaneous switching activity, an electrical simulation was performed using the Analog Artist simulator in the Cadence DFW2 environment. The circuits were mapped to the 0.8 μm CMOS technology from AMS. The results are shown in Figure 8.4. The electrical simulation transient is a fraction of the digital simulation from 0 to 1000 ns. The figures show the input pattern, the supply current waveform and some of the output signals. The correlation between the current waveform and the simultaneous switching activity obtained by the digital simulation is fairly good.

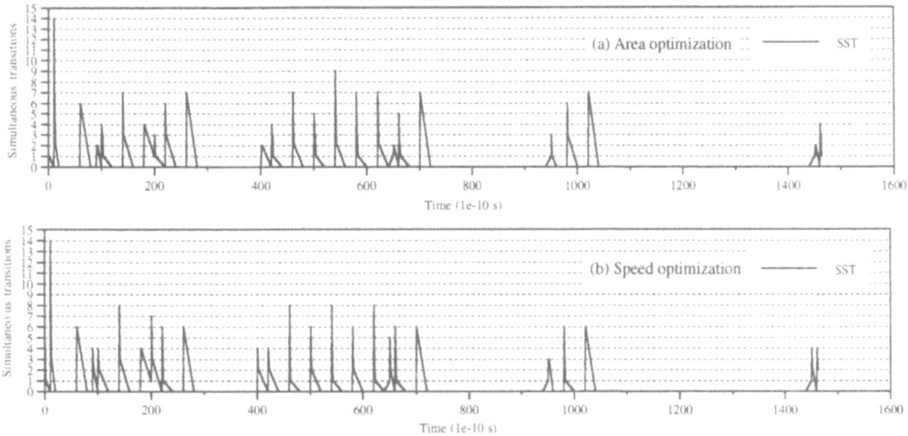

Figure 8.3

Simultaneous switching activity for the two synthesised solutions of the TLC circuit for the same input pattern obtained by digital simulation with Verilog.

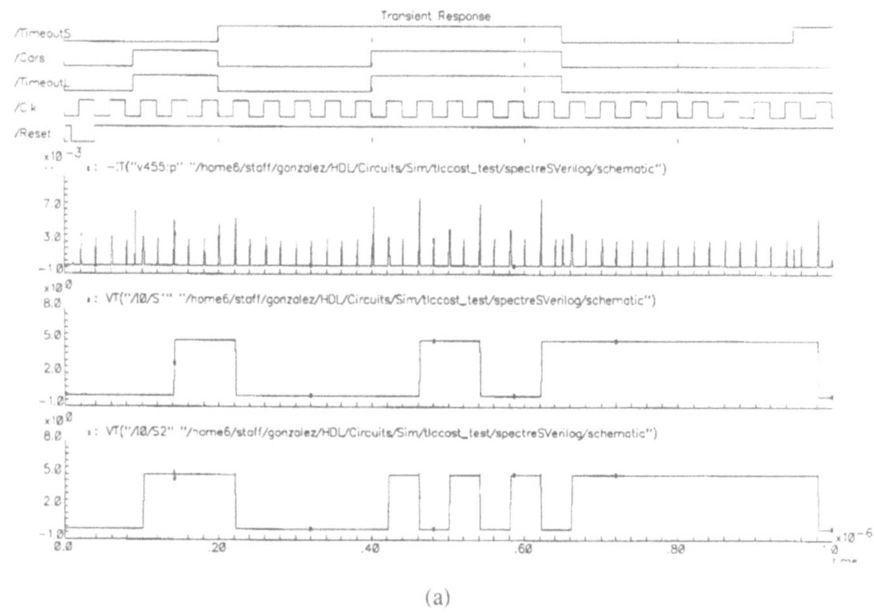

(a)

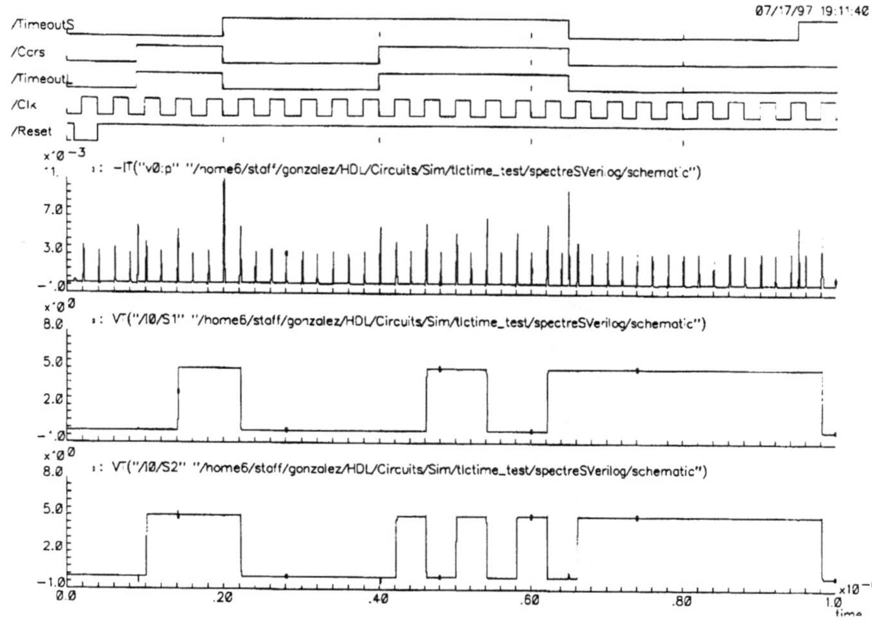

(b)

Figure 8.4

Electrical simulation results for the (a) TLCcost and (b) TLCtime circuits from the Analog Artist.

8.2.3 Example 2: 8 bit binary up/down counter

This second example is used to analyze a more complex circuit than the one in the previous example. The circuit is an 8 bit binary up/down counter structurally described using VHDL. The input file is shown in Figure 8.5 and Figure 8.6. The steps are the same as in the previous example. The two synthesized solutions with area and speed constraints are called COUNTcost and COUNTtime, respectively. Table 8.II shows the synthesis results. The simultaneous switching activity for the two solutions is shown in Figure 8.7. This example clearly shows the lower switching activity obtained with the minimum cost solution as compared with the maximum speed solution.

```
library ieee;
use ieee.std_logic_1164.all;

library work;
```

```
entity count_8 is
port(
        t: in std_logic; -- counter enable input
        clearn: in std_logic; -- comparator reset
        clk: in std_logic; -- clock
        q: out std_logic_vector(7 downto 0); -- output
        qn: out std_logic_vector(7 downto 0); -- negated output
       updown: in std_logic; -- direction control
        carry: out std_logic -- end of count
     );
end count_8;

architecture estructural of count_8 is

        component tffud -- counter basic cell
        port(
                t: in std_logic; -- T input
                clearn: in std_logic;     -- low-active reset
                clk: in std_logic;        -- clock
                v: out std_logic;-- for the next tffud
                q: out std_logic;-- output
                qn: out std_logic;        -- negated output
                updown: in std_logic      -- direction control
            );
        end component;

        signal carry_aux: std_logic_vector(6 downto 0);
        begin
           compta_bloc: for i in 7 downto 0 generate
                tff_0: if i = 0 generate
                gen_0: block
                       for all: tffud use entity work.tffud(behavioral);
               begin
                       tff_biest_0: tffud
                       port map
                       (
                       t => t,
                         clearn => clearn,
                         clk => clk,
                        v => carry_aux(0),
                         q => q(0),
                         qn => qn(0),
                         updown => updown
                       );
              end block;
              end generate;

                tff_7: if i = 7 generate
                gen_7: block
                       for all: tffud use entity work.tffud(behavioral);
               begin
                       tff_biest_7: tffud
                       port map
                       (
                       t => carry_aux(6),
                         clearn => clearn,
                         clk => clk,
                        v => carry,
                         q => q(7),
                         qn => qn(7),
                         updown => updown
                       );
              end block;
              end generate;

                tff_1_6: if i /= 0 and i /= 7 generate
                gen_1_6: block
                       for all: tffud use entity work.tffud(behavioral);
               begin
                       tff_biest_1_6: tffud
                       port map
                       (
```

```
                        t => carry_aux(i-1),
                        clearn => clearn,
                        clk => clk,
                        v => carry_aux(i),
                        q => q(i),
                        qn => qn(i),
                        updown => updown
                       );
                end block;
                end generate;
            end generate;
end estructural;
```

Figure 8.5

VHDL description of the 8 bit binary up/down counter.

```
library ieee;
use ieee.std_logic_1164.all;

entity tffud is

        port(
                t: in std_logic; -- input
                clearn: in std_logic;       -- low-active reset
                clk: in std_logic;          -- clock
                v: out std_logic;-- for the next tffud
                q: out std_logic;-- output
                qn: out std_logic;          -- negated output
                updown: in std_logic        -- direction control
            );
end tffud;

architecture behavioral of tffud is
        signal q_aux: std_logic; -- output auxiliary signal
        signal o_oxor: std_logic; -- conection xor_and
begin
        t1: process
        -- waits for the clock edge and checks reset (sincronous)
        begin
                wait until (clk'event and clk='1');
                if (clearn = '0') then
                        q_aux <= '0';
                elsif (t = '1') then
                        q_aux <= not q_aux;
                end if;
        end process t1;

        -- output assignment

        q <= q_aux;
        qn <= not q_aux;

        -- output generation for the next flip-flop

        o_oxor <= q_aux xor updown;
        v <= t and o_oxor;
end behavioral;
```

Figure 8.6

Behavioural description of the basic cell (tffud) used in the 8 bit binary up/down counter description.

Table 8.II.

Synthesis of results for the two solutions, COUNTcost and COUNTtime.

Parameter	COUNTcost	COUNTtime
Number of cells	62	166
Area (μm^2)	40601	130125
Maximum frequency	122.257 MHz	201.355 MHz
Longest path delay	8.179 ns	4.966 ns

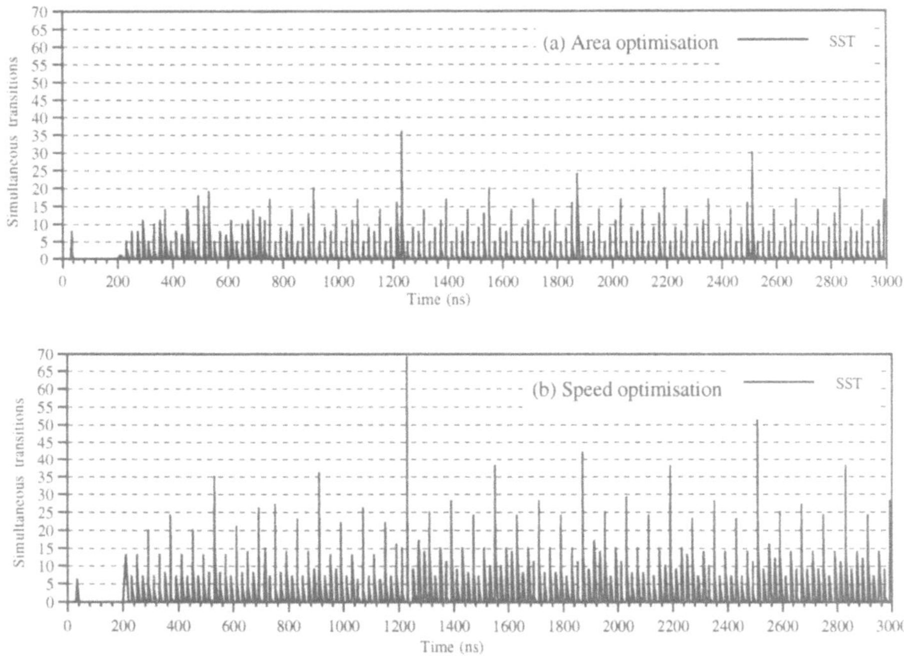

Figure 8.7

Simultaneous switching activity for the two synthesised solutions of the 8 bit binary up/down counter.

8.2.4 Example 3: numerical resolution of a differential equation

This example is widely used in literature to compare synthesis approaches. The circuit is a small computational loop that solves the following differential equation [6]:

$$\frac{d^2y}{dx^2}+3x\frac{dy}{dx}+3y=0$$

The variable u represents the dy/dx. dx is approximated by x_1-x and dy by y_1-y . Also du is computed as u_1-u. Parameter a is used to control the number of times the loop is executed by comparing x_1 and a. The operations performed in each loop are:

$$x_1 = x + dx$$

$$y_1 = y + udx$$

$$u = u - 3xudx - 3ydx$$

$$x = x_1 \qquad y = y_1$$

Figure 8.8 shows the VHDL description file for this circuit, called DIFFEQ. The synthesis process implies solving the trade-off between area and delay. The trade-off curve for the DIFFEQ circuit is shown in Figure 8.9. The structures for two of the trade-off points are shown in Figure 8.10 and Figure 8.11. The minimum delay solution has many blocks working in parallel and the simultaneous switching activity is greater than that of the minimum area solution.

```
-- Differential Equation Benchmark
--
-- Source: Adapted from example in paper
--        "HAL: A Multi-Paradigm Approach to Automatic Data Path Synthesis"
--         by P. Paulin, J. Knight and E. Girczyc
--         23rd DAC, June 1986, pp. 263-270
--
-- Benchmark author: Joe Lis
--
--    Copyright (c) 1989 by Joe Lis
--
-- Modified by Champaka Ramachandran on Aug 17th 1992
--
-- Verification Information:
--
--                   Verified     By whom?               Date          Simulator
--                   --------   ------------           --------      ------------
--  Syntax             yes    Champaka Ramachandran  17th Aug 92     ZYCAD
--  Functionality      yes    Champaka Ramachandran  17th Aug 92     ZYCAD
--------------------------------------------------------------------------------
library ieee;
use ieee.std_logic_1164.all;
use ieee.std_logic_arith.all;
library work;
use work.all;

entity diffeq is
   port (Xinport: in std_logic_vector(15 downto 0);
         Xoutport: out std_logic_vector(15 downto 0);
         DXport: in std_logic_vector(15 downto 0);
         Aport: in std_logic_vector(15 downto 0);
         Yinport: in std_logic_vector(15 downto 0);
         Youtport: out std_logic_vector(15 downto 0);
         Uinport: in std_logic_vector(15 downto 0);
         Uoutport: out std_logic_vector(15 downto 0);
          clk: in std_logic);
end diffeq;
```

```
--VSS: design_style BEHAVIORAL
architecture diffeq of diffeq is

begin
P1 : process (Aport, DXport, Xinport, Yinport, Uinport,clk)
    variable x_var,y_var,a_var,dx_var,u_var,x1,y1,t1,t2,t3,t4,t5,t6,t7:
std_logic_vector(15 downto 0);

begin
if (clk'event and clk='1') then
  x_var := Xinport;
  a_var := Aport;
  dx_var:= DXport;
  y_var := Yinport;
  u_var := Uinport;

   if (a_var > x1) then
        x1 := x_var + dx_var;
        t1 := u_var * dx_var;
        y1 := y_var + t1;
        t2 := "0000000000000011" * x_var;
        t3 := "0000000000000011" * y_var;
        t4 := t1 * t2;
        t5 := dx_var * t3;
        t6 := u_var - t4;
        u_var := t6 - t5;
        x_var := x1;
        y_var := y1;
   end if;
  end if;

  Xoutport <= x_var;
  Youtport <= y_var;
  Uoutport <= u_var;
end process P1;

end diffeq;
```

Figure 8.8

Behavioural VHDL description for the DIFFEQ circuit.

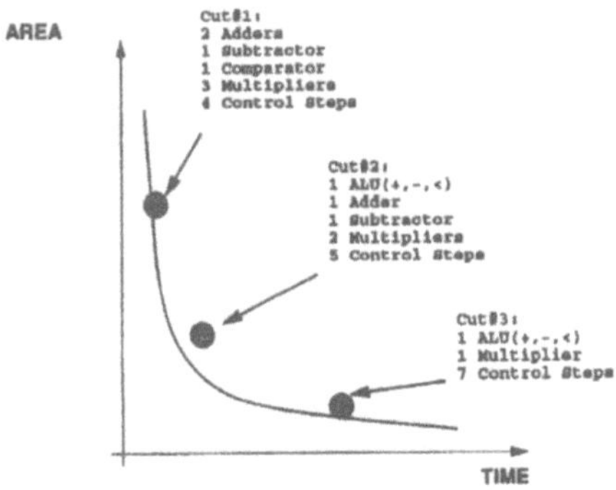

Figure 8.9

Trade-off curve for the DIFFEQ circuit synthesis process [7].

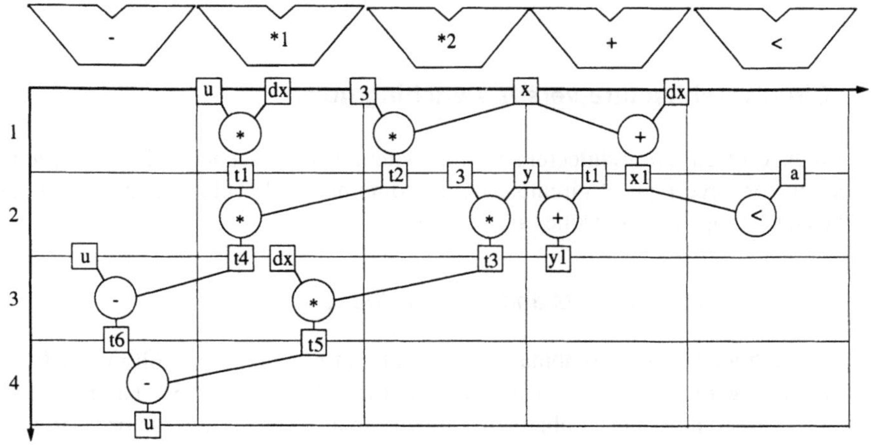

Figure 8.10

Structure and timing representation of the trade-off curve central point for the DIFFEQ circuit

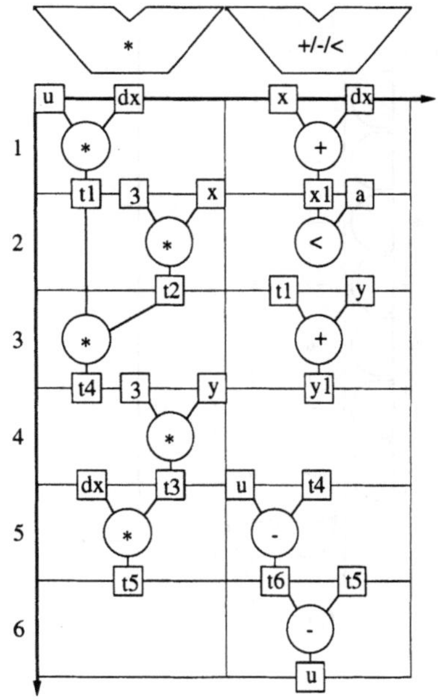

Figure 8.11

Minimum area solution for the DIFFEQ circuit.

8.3 Circuit architecture *versus* switching noise

The influence of circuit architecture in the switching noise generated is analyzed in this section for three type of circuits: purely combinational logic circuits, finite state machines and datapath + controller architectures.

8.3.1 Combinational circuits and logic synthesis

The basic architecture of a combinational circuit can be one of the following: ROM-based, PLA (Figure 8.12), selector-based or multilevel (Figure 8.13). The best trade-off between area and delay is obtained with the multilevel logic structure. However the PLA logic circuits are easier to synthesize and are commonly found in many circuit designs [8].

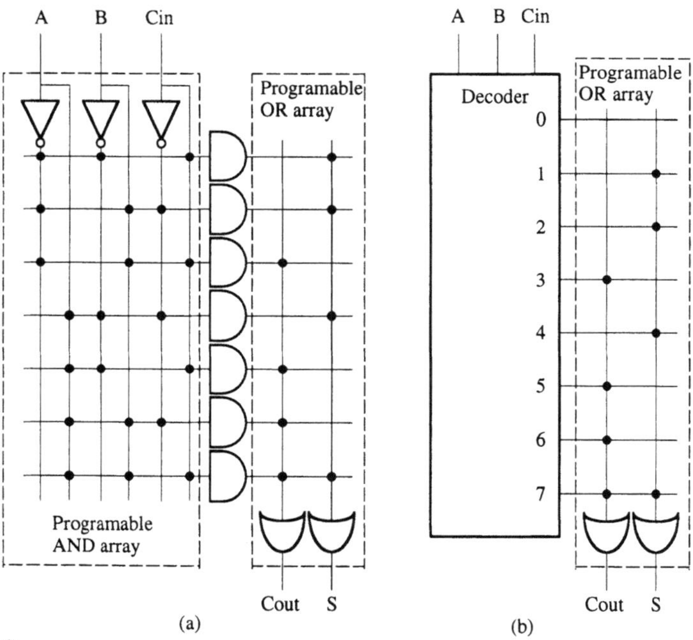

Figure 8.12

Full adder implementation using (a) a PLA and (b) a ROM.

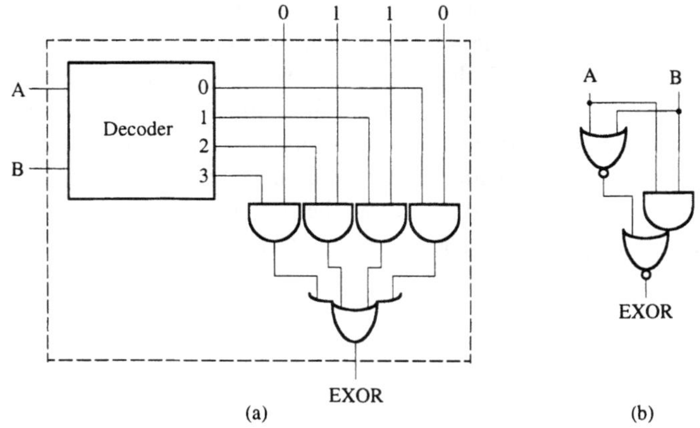

Figure 8.13

EXOR implementation using (a) a selector and (b) multilevel logic gates.

The regular two-level structure of the PLA based logic circuits is more prone to simultaneous switching noise. However it is not clear that all the product terms in the first level of the PLA structure would switch at the same time. This depends on the logic function, as is illustrated in Figure 8.14, where different covers for the same logic function are shown. The minimum switching noise is obtained with cover f_c that uses cubes without intersections between them. In the cover of f_a all the four product terms or cubes have a transition for some pairs of input vectors. In the cover of f_c the minimum number of simultaneous transitions is obtained at the expense of entailing a higher number of variables in two of the four cubes. This logic synthesis strategy can be taken on to more complex logic functions, even including "don't cares".

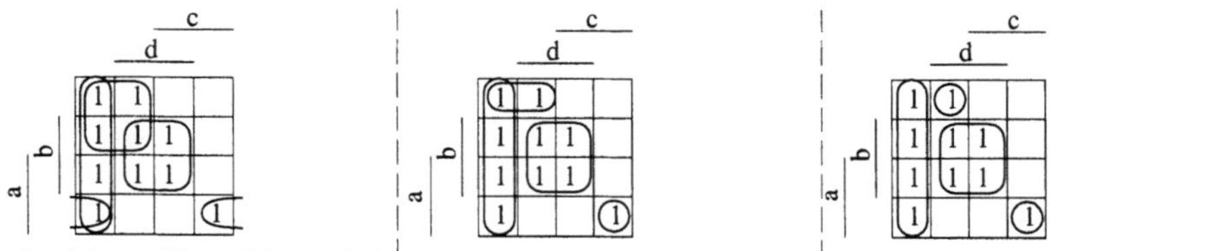

f_a= a'c'	+ c'd'	+ bd	+ ab'd'
0000	0000	0101	1000
0001	0100	0111	1010
0100	1000	1101	
0101	1100	1111	

f_b= a'b'c'	+ c'd'	+ bd	+ ab'cd'
0000	0000	0101	1100
0001	0100	0111	1110
	1000	1101	
	1100	1111	

f_c= a'b'c'd	+ c'd'	+ bd	+ ab'cd'
0001	0000	0101	1100
	0100	0111	1110
	1000	1101	
	1100	1111	

Figure 8.14

Different covering alternatives for the same logic function.

The problem of minimizing simultaneous switching activity in multilevel logic structures is more complex. The influence of the different propagation paths and the different delays associated with these rules out a simple approach. Strategies such as retiming the circuit can be used to alleviate the number of simultaneously switching events at a given time. Nevertheless, glitches, process variations, and the influence of interconnection-related propagation delay accentuate the complexity of the problem. A simple study was performed to show the influence of the logic structure on simultaneous switching activity. The same logic function is implemented with two

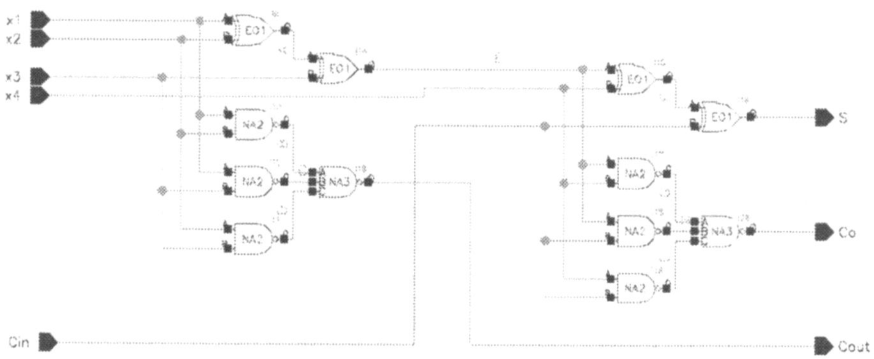

Figure 8.15

2xfa circuit schematic.

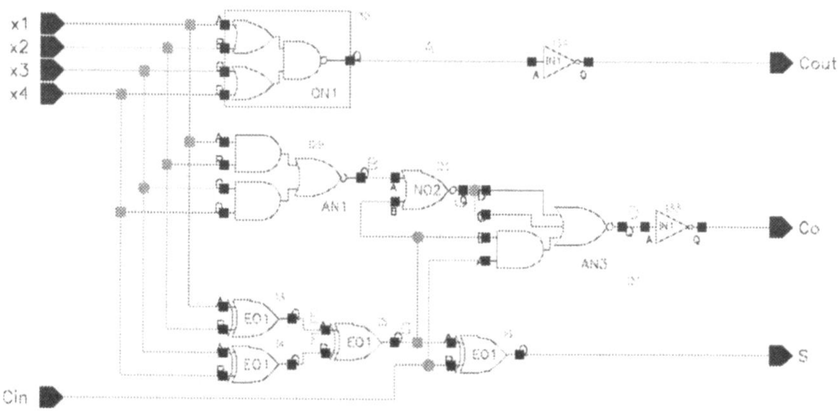

Figure 8.16

Compressor circuit schematic.

different logic circuits. The first is a cascade connection of two full adders (this circuit is labeled *2xfa*) and is shown in Figure 8.15. The second is a simpler version of the circuit labeled *compressor* and is shown in Figure 8.16. The circuits are implemented using gates from the standard cell library of 0.8 μm AMS CMOS technology. The logic simulation performed by Verilog reveals higher simultaneous activity of the *2xfa* circuit, as shown in Figure 8.17.

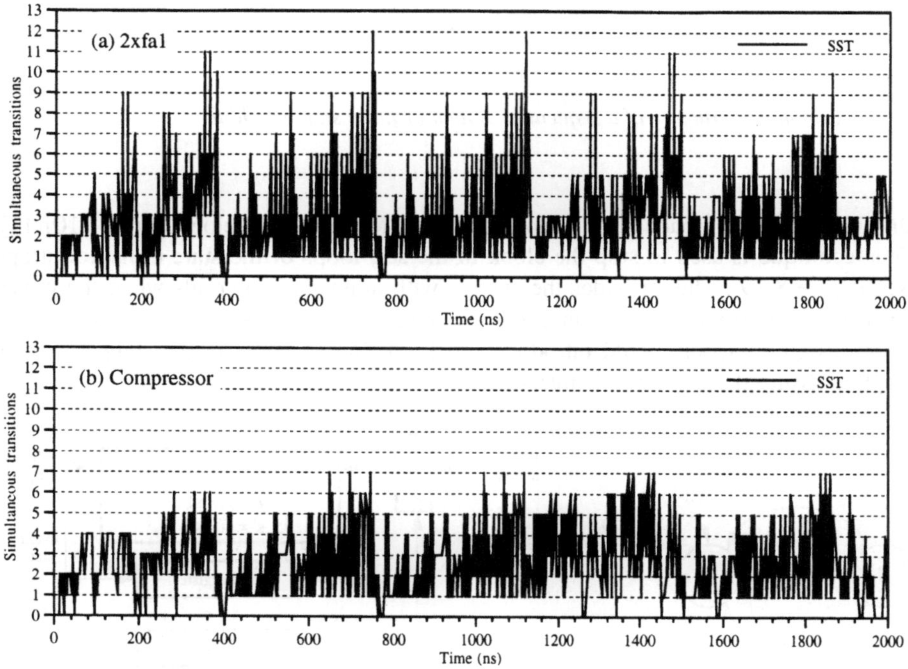

Figure 8.17

Simulation results for the 2xfa and compressor circuits showing the simultaneous switching activity.

8.3.2 Finite state machines

The simpler sequential circuits, finite state machines (FSM), have a fixed canonical structure (see Figure 8.18). The amount of switching noise generated by the pure sequential part of the structure (the flip-flops) depends on the number of bits used to encode the state of the system that switch simultaneously. Proper state encoding would reduce the switching noise by assigning adjacent codes to the states [9].

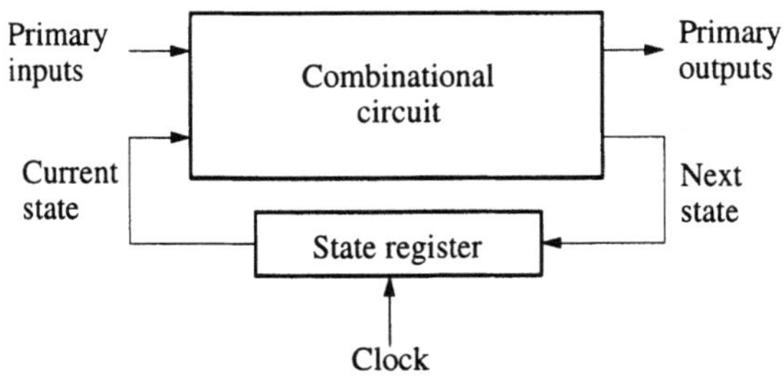

Figure 8.18

Canonical structure of a sequential system or finite state machine.

One interesting approach used in low power FSM design is clock gating [10][11]. This technique is used to stop the clock in the self-loops of the state transition graph (see Figure 8.19). This provides the system with quiet time intervals when the noise-sensitive parts of the total circuit could be enabled. A complementary low switching noise design approach would introduce self-loops intentionally in the state transition graph to assure these quiet time intervals.

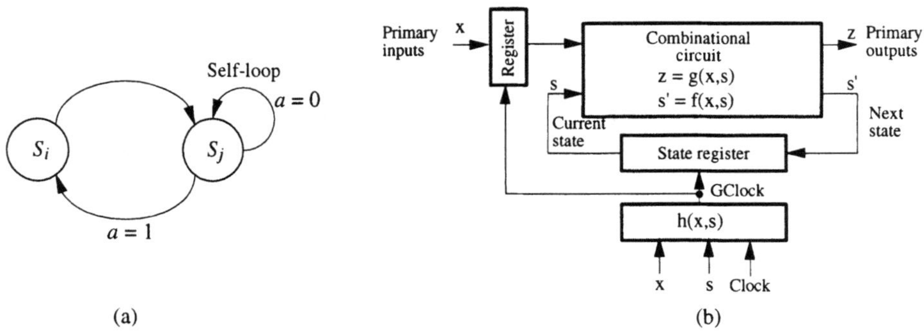

Figure 8.19

a) State transition diagram including a self-loop in state S_j. (b) Modified FSM structure including the clock gating circuit.

8.3.3 Datapath and control unit structure

With complex circuits the structure is divided into a datapath and a control unit. The control unit is implemented with an FSM. The datapath is made up of registers, operators (such as multipliers, ALUs, etc.) and multiplexers. The datapath structure influences the switching noise generated through the simultaneous switching activity that could potentially be found in the circuit. This is connected with how much concurrence there is in the datapath. The main objective of datapath designers is to maximize the throughput (number of operations per time unit). The structure of the datapath can be modified to improve the throughput in two ways. One is to increase the number of resources and make them work in parallel, as illustrated in the previous section with the DIFFEQ example. The other is to transform the datapath into a pipeline structure. Both strategies increase the number of simultaneous switching events and are therefore not desirable from the low switching noise point of view. However, a trade-off could be found between the throughput and the maximum simultaneous switching activity allowed that would drive the datapath design. The trade-off will limit the maximum number of parallel blocks or segments in a pipeline.

8.4 Logic alternatives *versus* switching noise

The logic alternative used to implement logic gates influences the performance of digital circuits in terms of speed, size, power consumption and routing complexity. In this section switching noise is used as the performance comparison parameter. The amount of switching noise generated by a given logic alternative involves two different aspects. The first is the driving capability of the gate needed to have the required switching times for typical output loads, that is in turn connected with the complexity of the gate structure. The second is connected with the way the gates are operated in a complete logic circuit (i.e. static CMOS logic circuits are inherently asynchronous and dynamic CMOS logic circuits are inherently synchronous). The logic alternatives were classified according to [12] and [13] as presented below. The technology considered is always the CMOS, when not expressly stated otherwise.

8.4.1 Static logics

The main characteristic of these logic alternatives is that a change in the input is instantaneously evaluated and propagated to the outputs. If two inputs of a gate inside the circuit have different arrival times there could be glitches at the output of this gate that are added to the normal switching activity of the logic circuit.

8.4.1.1 Conventional CMOS logics.

Complementary static CMOS logic gates will be used as reference logic alternative for the comparison. These have excellent noise margins. However, PMOS networks introduce high output loads because they tend to be 2 to 3 times bigger than the NMOS. For the same transition time requirement, the driving capability of these gates must be high compared with other logic alternatives.

8.4.1.2 Pseudo-NMOS logic.

The main advantage of this logic alternative is the simplicity of the gate structure that now has a single NMOS pull-down network. The lower output load relaxes the driving requirement and allows lower switching current pulses than conventional CMOS gates for the same transition times, and thus lower levels of switching noise. Its noise margins are nevertheless smaller because the PMOS pull-up limits the low logic level.

8.4.1.3 Pass transistor logic.

In recent years this logic has been proposed for low power and high speed applications (i.e. the CPL logic of [14]). Its main characteristic is using the drains of transistors, along with the gates, as logic inputs. Only a PMOS or NMOS network of pass transistors is sufficient to implement the logic function. The disadvantage is the voltage drop imposed by the pass transistors in one of the two logic levels. Ending buffers that dissipate static power must restore the levels. The noise margins are smaller than conventional CMOS. A proposed improvement put forward in [15] (double pass transistor logic or DPL) has the disadvantage of higher output loads due to the complementary structure of the pass transistor networks.

8.4.2 Non-pipeline dynamic logics

Dynamic CMOS circuits are used in high-speed applications. These circuits have the same advantages as Pseudo-NMOS logics but with a better power-delay trade-off. The main disadvantage is including a clock signal that controls gate operation. In one of the half cycles of the clock signal all the gates of the circuit perform a precharge operation. This represents a large number of simultaneous transitions. The switching noise generated by a single gate is smaller than the conventional CMOS, due to the simpler gate structure, but the switching noise generated by the whole circuit is much higher [16].

8.4.2.1 Domino logic and NP Domino logic.

Domino logic consists only of NMOS blocks [17]. It is very fast, but static CMOS inverters must be inserted between two domino gates. NP Domino logic uses alternate NMOS and PMOS blocks, eliminating the static inverter [18]. It is slower due to the PMOS transistors and needs two clock phases. This doubles the clock distribution network and increases the switching noise generated. To solve the charge-sharing problem during the evaluation phase Zipper Domino logic was proposed [19]. This needs four clock phases and worsens switching noise generation even further.

8.4.2.2 Differential Cascade Voltage Switch Logic (DCVSL).

Dynamic logic structures are transformed into static differential logics by including feedback transistors and complementary evaluation trees [20]. The number of nodes is doubled and one of the two complementary outputs always makes a transition in the evaluation phase. This is a disadvantage from the switching noise standpoint.

8.4.2.3 Differential Split-Level logic (DSL).

This is an interesting logic alternative, derived from DCVSL logic. The output nodes have a reduced swing of approximately $V_{DD}/2$ [21]. DSL gates can be cascaded by transmitting reduced swing logic signals. The final outputs restore full swing logic levels. Due to the lower voltage swings the switching current pulses are smaller and the same thing applies for switching noise. However, DSL gates have static power consumption and lower noise margins. The differential structure of the gates may compensate for the latter drawback.

8.4.3 Pipeline dynamic logics

One of the problems of dynamic logics is the signal race. The different arrival times of the inputs of a dynamic gate may erroneously discharge the dynamic output node, leading to a permanent error. For static gates this phenomenon is not permanent and is known as glitching. To solve this problem memory elements are inserted at the outputs of a chain of some dynamic gates, this structure leading to a pipeline of dynamic logic blocks. This approach is highly inadvisable for low switching noise design due to the increase in simultaneously switching blocks.

The best-known pipeline dynamic logics are the C^2MOS (Clocked CMOS) [22] and NORA (NO RAce) [23] logics, that use two clock phases. The TSPC (True Single Phase Clock) [24] and ANL (All N Logic) [13] logics have a single clock phase.

8.4.4 Differential clocked CMOS logics

This type of logic alternatives includes a latching circuit in the gate structure. They are derived from DCVS logic and have been proposed to improve some aspects of this form of logic (static power consumption, asymmetry of the output levels, large pull-down times [21]). The SSDL (Sample-Set Differential Logic) [25] uses a sense amplifier to accelerate gate switching, but involves the disadvantage of static power consumption. The ECDL (Enable/disable CMOS Differential Logic) solves that problem. All these logics have smaller transistor counts than the conventional static CMOS [26][27] to perform the same logic function, and are also faster, due to the lower loading obtained when using a single N or P evaluation block (not complementary structure) [28].

Due to the differential structure of the gates, one of the two outputs always performs a transition during the evaluation phase. The disable phase (equivalent to the precharge phase in dynamic logics) imposes a switching transition that is not found in static logics. These are disadvantages from the switching noise point of view. There is, however, an interesting application of this type of logics in self-timed circuits. A gate can generate the clock signal for the next gate when its operation is completed. 'Waves' of data are successively processed by the self-timed pipeline. This eliminates most of the simultaneous transitions because there is no global clock signal and the circuit smoothly computes the result. This type of circuit is known as wave pipeline [29][30] and will be used in the next chapter in a new low noise design technique.

8.4.5 *Synchronous design* versus *asynchronous design*

Over the last few decades asynchronous circuits have been proposed as a faster alternative to sequential circuits. The difficulty of designing these circuits and the sensitivity of the delay to the process and the layout have nevertheless constituted drawbacks preventing their widespread use. Asynchronous systems do not have a clock signal and the transitions can take place at any time, reducing the probability of simultaneous switching events. Asynchronous circuits are an optimum alternative for low noise design [31].

8.5 Low power design techniques and switching noise

System level low power techniques usually consist in turning off some parts of the circuit when they are not needed. These power-down strategies are very harmful from the low switching noise point of view because when the power-down modes are activated or de-activated the circuit demands large amounts of current. This problem has induced system integrators to change power supply distribution approaches [32][33][34].

The power consumed by a circuit can be expressed as $P_d = \alpha \cdot C_L \cdot V \cdot V_{DD} \cdot f_{CLK}$. where α is the activity factor, C_L the output load, V the output swing, that may be different from the power supply voltage (V_{DD}), and f_{CLK} is the clock frequency. Low power techniques at circuit level aim to reduce one of the several factors of the previous expression.

Reduction of the supply voltage has a quadratic effect on the reduction of power consumption but it also increases the delay. It furthermore reduces the switching noise generated by single gates, but noise margins are reduced too. To balance the delay increase produced by the supply voltage reduction, some authors suggest increasing parallelism or using pipeline structures [35]. Both techniques increase simultaneous activity and noise. There is a trade-off between low voltage-induced switching noise reduction and the increase in simultaneous activity that has to be dealt with.

Low power techniques on the algorithmic level focus on reducing circuit activity by exploiting some structural properties [36][37] or using optimal number systems [38].

The result of these techniques is a reduction in the total activity of the final circuit. However this does not necessarily imply that the maximum simultaneous activity is reduced by the same amount. Low power logic synthesis approaches [39][40][41][42] are used to reduce the total activity of synthesized circuits. What was stated above applies to these low power techniques.

To sum up, it is not clear that all low power techniques necessarily reduce switching noise. Some of them could even increase this (like power-down modes). The reduction of circuit activity does not necessarily reduce maximum simultaneous activity by the same amount, and this question indeed requires further analysis.

8.6 Low noise output driver design techniques

Output drivers of integrated circuits are considered to be one of the parts generating most switching noise, since they must commute large output loads in short times, and there is furthermore a possibility of all the outputs that form part of a bus switching simultaneously. Reduction techniques based on the use of on-chip decoupling capacitors are not useful in this case since the current loop always partly flows externally off the package. This means that the reduction of generated switching noise has great importance in the design of these output circuits. As the output drivers communicate several ICs, the noise generated in one of them might well affect the others, which is why it is also important to limit the amount of noise that can be propagated toward foreign circuits and systems, that may not necessarily have been designed to tolerate the same noise levels as the generating circuit. This section summarizes the different alternatives found in literature for reducing the noise of output drivers.

The first works on switching noise concentrated on the specifications of output drivers [43][44]. Later more accurate analyses were performed centering on CMOS technology [45]. The design of output drivers is basically determined by the output levels and required switching times. These switching times are designed for a maximum capacitive output load. As the off-chip loads are quite high (around tens of pF), the current pulses necessary to load and unload them are of great amplitude and the transition times demanded in high performance circuits are reduced (to around ns). A transition of 5 V amplitude on a capacity of 50 pF in 5 ns, assuming a current pulse in the form of an isosceles triangle, generates a dI/dt of 40 A/μs. This current pulse, across an inductance of 1 nH, becomes a 40 mV peak noise pulse. If, instead of a single driver, several drivers switch simultaneously in the same direction, for example 16, and assuming linear overlapping, the noise will be 640 mV. These numbers give an idea of what the problem involves.

The techniques used to reduce switching noise of output drivers, as in the case of the noise generated internally in the circuit, consist in reducing the inductance or reducing the *dI/dt*. The use of decoupling on-chip capacitors is not effective for the output drivers. Apart from this the current return path through the printed circuit board has also to be considered in the inductive factor (or that of common substrate, if the connection is between two chips within a MCM). So, not only is it important to design the package properly so as to reduce the inductance of the terminals, but also to assure low inductance board return paths.

Output drivers tend to consist of several cascaded inverters increasing in size from the input to the output of the driver, in what is known as tapered buffers. Selection of the number of stages and the sizes of each stage depends on the performance demanded of the driver and an optimum commitment between area and minimal propagation delay can be found [46]. The optimum design of these structures is affected by switching noise due to the fact that switching noise affects the delay of the switchings, as has been shown in chapter 2 [47]. Some previous works concentrated on optimizing the last stage of the tapered buffer and on the number of supply terminals necessary to obtain a given maximum switching noise limit, using long channel models [48] and also short channel models, which are appropriate for submicronic technologies to avoid implementations that turn out to be too conservative [49]. The problem of all these designs is that they are made for particular conditions that may later change. If the output load of these drivers turns out to be smaller, once the IC is mounted on the board, the driver will be oversized. A smaller load does not affect noise amplitude. In the best case it may reduce it (see the critical capacity concept in [45] and in chapter 7 of this book). Still, the structure of the RLC circuit formed by the driver, the package and the load is modified. The ringing at the output may therefore increase. Process variations also modify the initially designed sizes of the transistors. All these problems can be solved, at the expense of increasing output driver complexity, by means of the structure presented in [50], based on configurable cells that are internally adapted to compensate for the variations in the process and output load and to consistently obtain the same performance.

The first alternative for reducing the simultaneous switching noise (SSN) of the output drivers consists in delaying the switching of one part of the bus in respect of the other. If the period of the resonant oscillation or ringing of the outputs is known the ΔT delay between one part of the bus and the other must be adjusted so that the second switching coincides with the first minimal of the oscillation due to the switching of the first part. The two oscillations are therefore added up in counter phase and cancel each other out [51]. Another of the techniques proposed in this last work consists in adding a resistance serially with the outputs that help to attenuate oscillation amplitudes. This additional resistance slows down the switching and is needed to solve the design trade-off found between noise and switching speed.

Since then several specific output driver design proposals have been put forward to reduce the amount of dI/dt noise generated or to abate ringing. These techniques are based on controlling the waveform of the output switchings. Two types of techniques are found: those which control the slope of the outputs (controlled slew rate output buffers) and those which perform a stepped output switching.

Within the first group, in [52] resistances in series with the NMOS and PMOS supply terminals in the last stage of the driver are used to reduce rising and falling times, negatively affecting the delay. In [53] the resistances are serially added to the gate transistor outputs and their value is controlled with an external voltage so that an optimum value of the control voltage can be found for each load capacity that eliminates ringing in the outputs due to switching. In [54] two techniques are presented. The first consists in limiting the current provided by the output drivers. The second, more sophisticated, uses several output transistors of increasing size in parallel, which each commute with a delay in respect of the others. A transition with an increasingly greater slope is formed while one, two or three transistors in parallel start conducting. These kinds of drivers are known as controlled slew rate output drivers (CSR). The problem of all these solutions is that they are designed for typical values of the output capacity and if these vary, the final performance of the drivers may be worse than expected. The proposal in [55] consists of introducing a feedback that adjusts the switching characteristics of the driver to adapt it to the particular values of the output load.

The second group of techniques uses a stepped output switching that can be performed in two ways. It is done simply by using two pairs of output transistors in parallel and delaying the switching of one pair in respect of the other [56][57], or by precharging the output at an intermediate level, in a first phase of the switching, and after some time the output switching is completed [58][59]. In [60] the same technique is used but with the aim of reducing the ringing in the receiver, taking into account the distortion introduced by the transmission line that connects the output driver with the receivers in other ICs. All these techniques, like those which control the slope of the output switchings, involve the disadvantage of increasing switching time in respect of a conventional output driver.

8.7 Conclusions

High-level decisions have a considerable influence on the switching noise generated by the circuits finally produced. The synthesis process can be approached to produce minimum area or maximum speed implementations for the same original specification. The former is best suited for low noise design. Low noise-driven synthesis criteria could be used in circuit creation. One possible approach consists in retiming the data-flow graph to minimize the number of operations performed in the same time interval without affecting the longest path of the circuit. This and other possible low noise, high level design approaches are proposed for future research work.

The structural alternative also has a great influence on switching noise. Two level log structures (i.e. PLAs) are prone to generating greater amounts of noise than multilevel logic structures. However low-noise two-level optimization approaches have been proposed to reduce the number of simultaneous switching gates based on minimizing the common variables between the selected cubes of the cover for the logic function. Different multilevel structures for the same logic function may generate different levels of switching noise. It will be of great use to find low-noise multilevel logic synthesis algorithms, and this is also proposed for future research work.

Low power techniques used to synthesize FSMs or clock gating techniques have been proved to be useful low noise design alternatives. The increase in concurrent use of pipelines in the datapath is extremely inadvisable from the low switching noise standpoint due to the large amount of simultaneous activity entailed.

In general, conventional static CMOS logic provides the best compromise between noise generation and noise margins. However, there are some other interesting alternatives like static pseudo-NMOS (low output loads), DSL (low output voltage swings) or self-timed clocked CMOS differential logics (wave pipeline structure).

Low power techniques used at different design levels could reduce switching noise, especially if power consumption is reduced through the reduction in circuit activity. However there is no clear direct relationship between the reduction in total activity and the reduction in maximum simultaneous activity. Some low power techniques (such as power-down modes) may even be detrimental from the low switching noise standpoint.

References

[1] Peter J. Ashenden. (1996). *The Designer Guide to VHDL.* 1st edition. San Francisco: Morgan Kaufmann Publishers.

[2] D.E. Thomas, P.R. Moorby. (1991). *The Verilog Hardware Description Language*. Boston: Kluwer Academic Publishers.

[3] *Hamburg VHDL Archive*, Available in the WWW, URL <http://tech-www.informatik.uni-hamburg.de/vhdl/vhdl.html>, University of Hamburg.

[4] Synergy Synthesis Tools, Cadence Opus DFWII release 9502.

[5] D.E. Thomas, P.R. Moorby (1991) *The Verilog Hardware Description Language*. Massachusetts: Kluwer Academic Publishers.

[6] Pierre G. Paulin, Hohn P. Knight, 'Force-Directed Scheduling for the Behavioral Synthesis of ASIC's', *IEEE Tr. on Computer-Aided Design*, Vol. 8, No. 6, June 1989, pp. 661–679.

[7] Sabine März. 'High-Level Synthesis' in *The Synthesis Approach to Digital System Design*, Petra Michel, Ultich Lauter, and Peter Duzy, Ed. (1992). Massachusetts: Kluwer Academic Publishers.

[8] Daniel Gajski, Nikil Dutt, Allen Wu, and Steve Lin. (1992). *High-Level Synthesis: Introduction to Chip and System Design*. Massachusetts: Kluwer Academic Publishers.

[9] Luca Benini, and Giovani De Micheli, 'State Assignment for Low Power Dissipation', *IEEE J. Solid-State Circuits*, Vol. 30, No. 3, March 1995, pp. 258–267.

[10] Luca Benini, and Giovani De Micheli, 'Automatic Synthesis of Low-Power Gated-Clock Finite-State Machines', *IEEE Tr. on Computer-Aided Design of Integrated Circuits and Systems*, Vol. 15, No. 6, June 1996, pp. 630–643.

[11] Manfred Koegst, Günter Franke, Steffen Rülke, and Klaus Feske, 'A Strategy for Low Power FSM-Design by Reducing Switching Activity', *Proc. of 7th Int. Workshop PATMOS'97*, Lauvain-la Neuve, Belgium, 8–10 September 1997, pp. 119–128.

[12] Neil H.E. Weste, Kamran Eshraghian, *Principles of CMOS VLSI Design. A systems perspective*. 2d. Edition (1994). Massachusetts: Addison-Wesley Publishing Company.

[13] Richard X.Gu, Khaled M. Sharaf, Mohamed I. Elmasry. *High-Performance Digital VLSI Circuit Design*. 1996. Massachusetts: Kluwer Academic Publishers.

[14] K. Yano, et al., '3.8-ns CMOS 16x16-b Multiplier Using Complementary Pass-Transistor Logic', *IEEE J. Solid-State Circuits*, Vol. SC-25, April 1990, pp. 388–395.

[15] M. Suzuki, et al., 'A 1.5-ns 32-b CMOS ALU Double Pass-Transistor Logic', *IEEE J. Solid-State Circuits*, Vol. SC-28, Nov. 1993, pp. 1145–1151.

[16] Patrik Larsson, and Christer Svensson, 'Noise in Digital Dynamic CMOS Circuits', *IEEE J. Solid-State Circuits*, Vol. 29, No. 6, June 1994, pp. 655–662.

[17] R.H. Krambeck, Charles M. Lee, and Hung-Fai Stephen Law, 'High-Speed Compact Circuits with CMOS', *IEEE J. Solid-State Circuits*, Vol. SC-17, No. 3, June 1981, pp. 614–619.

[18] V. Friedman and S. Liu, 'Dynamic Logic CMOS Circuits', *IEEE J. Solid-State Circuits*, Vol. SC-19, April 1984, pp. 263–266.

[19] C.M. Lee, and E. W. Szeto, 'Zipper CMOS', *IEEE Circuits & Devices Mag.*, May 1986, pp. 10–16.

[20] Kan M. Chu, and David L. Pulfrey, 'A comparison of CMOS Circuit Techniques: Differential Cascode Voltage Switch Logic Versus Conventional Logic', *IEEE J. Solid-State Circuits*, Vol. SC-22, No. 4, August 1987, pp. 528–532.

[21] Leo C.M.G. Pfennings, Wim G.J. Mol, Joseph J.J. Bastiaens, and Jan M.F. Van Dijk, 'Differential Split-Level CMOS Logic for Subnanosecond Speeds', *IEEE J. Solid-State Circuits*, Vol. SC-20, No. 5, October 1985, pp. 1050–1055.

[22] Yasoji Susuki, Kaicho Odagawa, and Toshio Abe, 'Clocked CMOS Calculator Circuitry', *IEEE J. Solid-State Circuits*, Vol. SC-8, No. 6, Dec. 1973, pp. 462–469.

[23] Nelson F. Goncalves, Hugo J. De Man, 'NORA: A Recefree Dynamic CMOS Technique for Pipelined Logic Structures', *IEEE J. Solid-State Circuits*, Vol. SC-18, No.3, June 1983, pp. 261–266.

[24] Jiren Yuan and Christer Svensson, 'High-Speed CMOS Circuit Technique', *IEEE J. Solid-State Circuits*, Vol. 24, No. 1, February 1989, pp. 62–70.

[25] Timothy A. Grotjohn, and Bernd Hoeflinger, 'Sample-Set Differential Logic (SSDL) for Complex High-Speed VLSI', *IEEE J. Solid-State Circuits*, Vol. SC-21, No. 2, April 1986, pp. 367–369.

[26] Kan M. Chu, and David L. Pulfrey, 'Design Procedures for Differential Cascade Voltage Switch Circuits', *IEEE J. Solid-State Circuits*, Vol. SC-21, No. 6, December 1986, pp. 1082–1087.

[27] Jordi Cortadella, 'Mapping BDD's Into DCVSL Gates (or Switch-Level Synthesis)', Research report of the Computer Architecture Dep't, Universitat Politècnica de Catalunya, August 1992.

[28] Shih-Lien L. Lu, and Milos D. Ercegovac, 'Evaluation of Two-Summand Adders Implemented in ECDL CMOS Differential Logic', *IEEE J. Solid-State Circuits*, Vol. 26, No. 8, August 1991, pp. 1152–1160.

[29] Wei-han Lien, and Wayne P. Burleson, 'Wave-Domino Logic: Theory and Application', *IEEE Tr. on Circuits and Systems—II*, Vol. 42, No. 2, February 1995, pp. 78–91.

[30] C.H. Lau, D. Renshaw, and J. Mavor, 'A Self-Timed Wavefront Array Multiplier', *Proc. of 1989 Int. Symp. on Circuits and Systems (ISCAS'89)*, pp. 138–141.

[31] Vitit Kantaburata, and Andreas G. Andreou, 'A State Assignment Approach to Asynchronous CMOS Circuit Design', *IEEE Tr. on Computers*, Vol. 43, No. 4, April 1994, pp. 460–469.

[32] Bob Mammano, 'Fuelling the Megaprocessors - Empowering Dynamic Energy Management', —.

[33] Rich Evans, and Michael Tsuk, 'Modelling and Measurement of High-Performance Computer Power Distribution System', *IEEE Tr. on Components, Packaging, and Manufacturing Tech.—Part B*, Vol. 17, No. 4, November 1994, pp. 467–471.

[34] *VRM 9.1 DC-DC Converter Design Guidelines*. Intel Application Notes, Order Number 243408-001. May 1997.

[35] Anantha P. Chandrakasan, Samuel Sheng, and Robert W. Brodersen, 'Low-Power CMOS Digital Design', *IEEE J. Solid-State Circuits*, Vol. 27, No. 4, April 1992, pp. 473–484.

[36] Anantha P. Chandrakasan, Miodrag Potkonjak, Renu Mehra, Jan Rabaey, and Robert W. Brodersen, 'Optimising Power Using Transformations', *IEEE Tr. on Computer-Aided Design of Integrated Circuits and Systems*, Vol. 14, No. 1, January 1995, pp. 12–31.

[37] Jan M. Rabaey, and Massoud Pedram, editors. *Low Power Design Methodologies*, 1996, Massachusetts: Kluwer Academic Publishers.

[38] Vasily G. Moshnyaga and Keikichi Tamaru, 'A Comparative Study of Switching Activity Reduction Techniques for Design of Low-Power Multipliers', *Proc. of 1995 IEEE Int. Symp. Circuits and Systems: ISCAS'95*, pp. 1560–1563.

[39] Anantha P. Chandrakasan, and Robert Brodersen. *Low Power Digital CMOS Design*. 1995. Massachusetts: Kluwer Academic Publishers.

[40] P. Girard, C. Landrault, S. Pravossoudovitch, and D. Severac, 'Technique for Reducing Power Consumption in CMOS Circuits', *Electronics Letters*, Vol. 33, No. 6, March 1997, pp. 485–486.

[41] R. Iris Bahar, Fabio Somenzi, 'Boolean Techniques for Low Power Driven Re-Synthesis', *Proc. of IEEE ICAD Conf.*, 1995, pp. 428–432.

[42] G. Theodoridis, S. Theoharis, D. Soudris, O. Koufopavlou, and C. Goutis, 'A Novel Approach for Reducing the Switching Activity in Two-Level Logic Circuits', *Proc. of ICECS*, 1996, pp. 840–843.

[43] E.E. Davidson, 'Electrical Design of a High Speed Computer Package', *IBM J. Res. Develop.*, Vol. 26, No. 3, May 1982, pp. 349–361.

[44] George A. Katopis, 'Delta-I Noise Specification for a High-Performance Computing Machine', *Proceedings of the IEEE*, Vol. 73, No. 9, September 1985, pp. 1405–1415.

[45] Arun Vaidyanath, Bigir Thoroddsen, and J.L. Prince, 'Effect of CMOS Driver Loading Conditions on Simultaneous Switching Noise', *IEEE Tr. on Components, Packaging, and Manufacturing Techn.—Part B*, Vol. 17, No. 4, November 1994, pp. 480–485,

[46] Hung Chang Lin, and Loren W Linholm, 'An Optimized Output Stage for MOS Integrated Circuits', *IEEE J. Solid-State Circuits*, Vol. SC-10, No. 2, April 1975, pp. 106–109.

[47] Srivasa R. Vemuru, 'Effects of Simultaneously Switching Noise on the Tapered Buffer Design', *IEEE Tr. on Very Large Scale Integration (VLSI) Systems*, Vol. 5, No. 3, September de 1997, pp. 290–300.

[48] Yaochao Yang, and John R. Brews, 'Design Trade-Offs for the Last Stage of an Unregulated, Long-Channel CMOS Off-Chip Driver with Simultaneous Switching Noise and Switching Time Considerations', *IEEE Tr. on Components, Packaging, and Manufacturing Techn.—Part B*, Vol. 19, No. 3, August 1996, pp. 481–486.

[49] Yahochao Yang and John R. Brews, 'Design for Velocity Saturated, Short-Channel CMOS Drivers with Simultaneous Switching Noise and Switching Time Considerations', *IEEE J. Solid-State Circuits*, Vol. 31, No. 9, September 1996, pp. 1357–1360.

[50] Thaddeus J. Gabara, Wilhelm C. Fischer, John Harrington, and William W. Troutman, 'Forming Damped LRC Parasitic Circuits in Simultaneously Switched CMOS Output Buffers', *IEEE J. Solid-State Circuits*, Vol. 32, No. 3, March. 1997, pp. 407–418.

[51] R. Senthinathan, J.L. Prince, and S. Nimmagadda, 'Noise Immunity Characteristics of CMOS Receivers and Effects of Skewing/Damping CMOS Driver Switching Waveform on the Simultaneous Switching Noise', *Microelectronics Journal*, Vol 23, 1992, pp. 29–36.

[52] Karl L. Wang, Mark. D. Bader, Vince W. Soorholtz, Richard W. Mauntel, Horacio J. Mendez, Peter H. Voss, and Roger I. Kung, 'A 21-ns 32K×8 CMOS Static RAM with a Selectively Pumped p-Well Array', *IEEE J. Solid-State Circuits*, Vol. SC-22, No. 5, October 1987, pp. 704–711.

[53] L. Alberston, S. Witaker and R. Merrell, 'CMOS Output Buffer Waveshaping', *Proc. of 1st Great Lakes Symp. on VLSI*, 1991, pp. 326–327.

[54] R. Senthinathan, and J.L. Prince, 'Application-Specific CMOS Output Driver CIrcuit Design Techniques to Reduce Simultaneous Switching Noise', *IEEE J. Solid-State Circuits*, Vol. 28, No. 12, December 1993, pp. 1383–1388.

[55] C.S. Choy, M.H. Ku, and C.F. Chan, 'A Low Power-Noise Output Driver with an Adaptative Characteristic Applicable to a Wide Range of Loading Conditions', *IEEE J. Solid-State Circuits*, Vol. 32, No. 6, June 1997, pp. 913–917.

[56] Donald T. Wong, R. Dean Adams, Arup Bhattacharyya, James Covino, John A. Gabric, and George M. Lattimore, 'An 11-ns 8k × 18 CMOS Static RAM with 0.5-μm Devices', *IEEE J. Solid-State Circuits*, Vol. 23, No. 5, October 1988, pp. 1095–1103.

[57] K. Furutani, H. Miyamoto, Y. Morooka, M. Suwa, and H. Ozaki, 'An Adjustable Output Driver with a Self-Recovering Vpp Generator for a 4M × 16 DRAM', *IEEE J. Solid-State Circuits*, Vol. 29, No. 3, March 1994, pp. 308–310.

[58] Tomohisa Wada, Toshihiko Hirose, Hirofumi Shinohara, Yuji Kawai, Kojiro Yuzuriha, Yoshiho Kohono, and Shimpei Kayano, 'A 34-ns 1-Mbit CMOS SRAM using triple Polysilicon', *IEEE J. Solid-State Circuits*, Vol. SC-22, No. 5, October 1987, pp. 727–732.

[59] Ernestina Chiofi, Franco Maloberti, Gianmarco Marchesi, and Guido Torelli, 'High-Speed, Low-Switching Noise CMOS Memory Data Output Buffer', *IEEE J. Solid-State Circuits*, Vol. 29, No. 11, November 1994, pp. 1359–1365.

[60] Tomonori Sekiguchi, Masashi Horiguchi, Takeshi Sakata, Yoshinobu Nakagome, Shigeki Ueda, and Masakazu Aoki, 'Low-Noise, High-Speed Data Transmission Using a Ringing-Canceling Output Buffer', *IEEE J. Solid-State Circuits*, Vol.30, No. 12, December 1995, pp. 1569–1574.

9 Techniques and circuits for reducing switching noise

This chapter presents a brief analysis of several known techniques and circuits used for reducing switching noise, along with some new proposals both for the power supply distribution side and digital circuit structure.

9.1 Introduction

Switching noise generation depends on two different factors: the inductances of the power supply distribution network and the time derivative of the whole circuit's current consumption waveform. Switching noise reduction can be tackled from one (or both) of these two angles. This chapter contains a description of several techniques and circuit solutions reported in the literature now available as well as some new proposals. Firstly low noise solutions based on current consumption waveform shaping are approached, and then solutions based on power supply network design.

9.2 Constant current consumption logics

The origin of constant current consumption logics lies in the world of emitter coupled logic (ECL). In CMOS technology these logics evolve into current steering logics (CSL) [1]. The current circulates alternately through one of the two symmetric halves of the circuit depending on the logic function. Nevertheless, from the power supply point of view, the current flow is constant, and thus switching noise is theoretically avoided. Several current steering logic alternatives will now be summarized. The main disadvantage of this kind of low noise solution is the static power consumption and non-standard logic levels.

9.2.1 Enhancement Source-Coupled Logic (ESCL)

The first proposal found in the literature is the ESCL [2][3]. Figure 9.1.a shows an ESCL inverter implemented using only NMOS transistors. The differential input stage formed by *m1* and *m2* is biased by a constant current *I*, and the loading NMOS (*m3* and *m4*) are biased by V_B to work in the saturation region. A high voltage at input *A* puts *m1* transistor in on and the low voltage at *A'* cuts off *m2*. The current flows only through *m1* and the differential output voltage generated is:

$$\Delta V_{out} = \left[2I_1 / k_n'(W/L)_3\right]^{1/2} \qquad Eq.\ (9.1)$$

where $(W/L)_3$ is the *m2* aspect ratio, and k_n' the NMOS unit current gain. The ECDL gates design is completed by setting a voltage gain (a_v) greater than one in order to get good noise margins:

$$a_v = (W/L)_1/(W/L)_3 \geq 1 \qquad Eq.\ (9.2)$$

where $(W/L)_1$ is the *m1* aspect ratio.

The main advantages of ESCL logic from CMOS are the reduction of switching current pulses in the power lines and substrate, smaller output voltage swing, and higher noise immunity. ESCL has differential output logic values that may simplify logic design. Any logic function can be implemented using an NMOS logic tree, as shown in Figure 9.1.b. The bias voltage V_B tends to be connected to V_{DD} to reduce the number of interconnections.

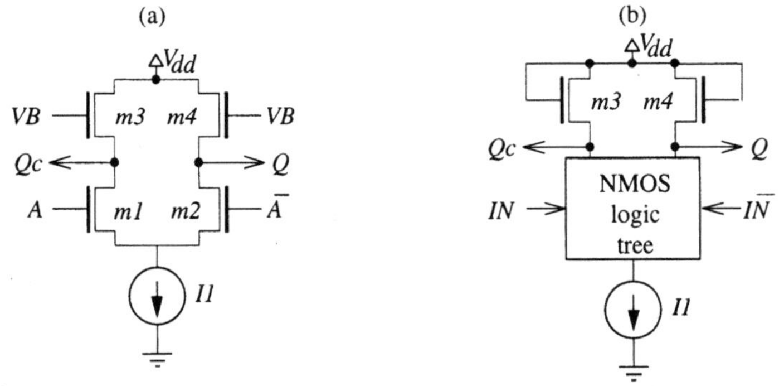

Figure 9.1

a) *ESCL inverter. (b) General ESCL gates structure.*

9.2.2 Folded Source-Coupled Logic (FSCL)

FSCL logic [4][5] is an extension of SCL logic [2] that improves its bandwidth and noise reduction performance. FSCL gates consist of a differential input stage (*m1* and *m2*) and an output stage implemented with two diode-connected PMOS that converts the current into voltage (see Figure 9.2). The output voltage swing is:

$$\Delta V_{out} = \left[2I_2 / k_p^{'}(W/L)_4\right]^{1/2} - \left[2(I2 - I1)/k_p^{'}(W/L)_3\right]^{1/2} \qquad Eq.\ (9.3)$$

This tends to range from 0.5V to 1 V. The voltage gain of the gate must be greater than two if good noise margins are to be obtained.

Any logic function can be implemented using a differential NMOS logic tree [6]. The FSCL switching noise generated is two orders of magnitude smaller than that of the functionally equivalent CMOS gates (see Figure 9.3).

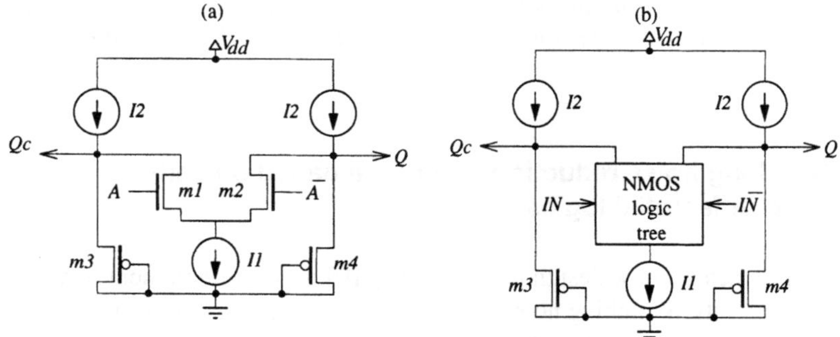

Figure 9.2

(a) FSCL inverter. (b) Generic FSCL gate structure.

9.2.3 Other CSL logic alternatives

Some authors have proposed alternatives to the previous logics in order to reduce transistor count, such as current-balanced logic (NMOS-CBL) [7], in which the evaluation NMOS tree is made up of one sole branch. CSL logic, as reported in [1], is a single-ended version of FSCL logic. However it is difficult to find any real applications that use all these current steering logics. [8] presents a mixed signal IC to perform arrival time measurements in a particle detector. In this circuit, the logic blocks are implemented with a bipolar current steering structure and the MOS transistors are used as current references and active loads as in BiCMOS technology.

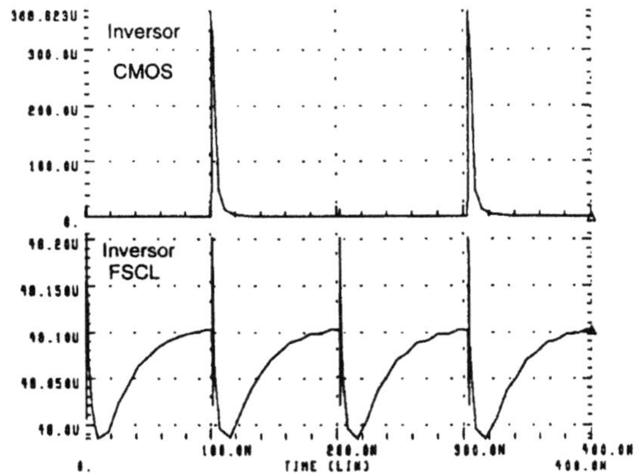

Figure 9.3

Comparison of the current peaks generated when switching a 0.25 pF output load by a static CMOS inverter (upper graph) and an FSCL inverter (lower graph).

9.3 Switching noise reduction technique based on self-timed differential logics

This section presents a new design technique [9][10] used to implement low switching noise digital circuits. Switching noise reduction is achieved by controlling the shape of the switching current waveform of the CMOS logic circuits. Self-timed structures are required to obtain the required switching current waveform shape. Current limiters are also used to control the current waveform amplitude of the structure's single cells. The design strategy proposed is applied to a circuit example, a 4x4 unsigned array multiplier, and experimental results are given.

Switching noise is generated by the inductive parasitics of package power supply terminals and on-chip power supply distribution when several digital cells switch simultaneously. Switching noise is proportional to the product of the effective inductance of the circuits' power supply terminals and the time derivative of the switching current waveform. Switching noise can be reduced by lowering this effective power supply inductance, as will be seen in the next section. It can be also reduced by minimizing the time derivative of the switching current waveform. The aim of the technique proposed is to obtain a switching current waveform with minimum time derivatives. The obvious solution is a constant current waveform, as shown in the previous section. Unfortunately, these current steering logics involve higher power consumption and non-full rail output swings. The alternative proposed in this section for the switching current waveform is a trapezium (Figure 9.4.a). The rising and falling edges of this current waveform, produced at the start and end of the logic circuitry operation, generate switching noise. However, during the operation of the circuit the flat part has a zero time derivative. Such a trapezium switching current waveform would be obtained by the overlapping of the single switching current waveforms of the circuit cells. It is important for the single cell's current waveforms to be controllable and for them to overlap properly.

9.3.1 Design methodology

A logic circuit is taken as an interconnection of basic logic gates or blocks. After the inputs change, different gates consecutively switch until the final result appears at the outputs. At a given time, one or more gates could switch simultaneously. The final current waveform is the temporal overlap of the individual gates' current waveform. The waveform shape depends on the individual current shape and amplitude of each gate, the number of simultaneous switching gates and the switching times

In order to obtain a trapezoidal current waveform, the number of gates switching simultaneously at a given time must always be the same. The shape and amplitude of the individual current waveforms must be the same for all the different logic gates, regardless of the input vector and the gate's logical function. This is obtained by following the design rules below:

1. Making the logic circuit as regular as possible, adding redundant blocks in order to have an equal number of switching gates at a given time for any input combination.
2. Using a logic that will produce switching current pulses of approximately the same shape regardless of the input changes.
3. The delay of each block or gate of the regular structure must be the same. If necessary big blocks can be partitioned into smaller ones until spatial and temporal regularity is achieved.

The above design conditions are not satisfied in a static CMOS circuit. The switching of a static CMOS gate depends on the input vector change (i.e. a change in the inputs of a static CMOS NAND from 00 to 11 produces a current spike, but a change from 00 to 01 does not). The current peak's amplitude also depends on the input vector (in a static CMOS NAND the current spike in the switching produced by a 00 to 11 is approximately twice as large as the one for a 01 to 11 input change).

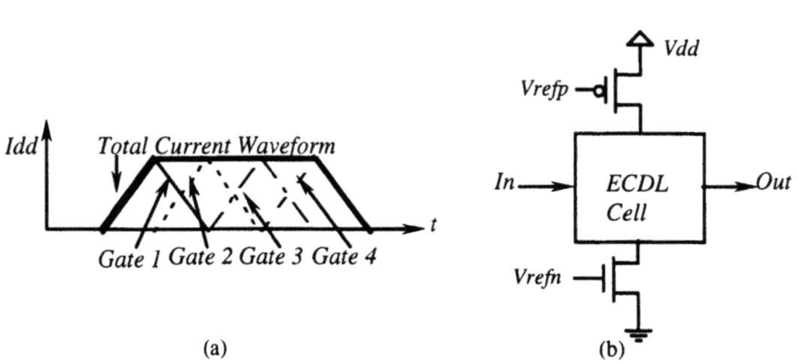

Figure 9.4

(a) Ideal trapezoidal current waveform obtained by overlapping four individual gate currents. (b) Current limiters in the power and ground nodes of an ECDL cell.

9.3.2 Low noise design example: 4 × 4 array multiplier

We base our design approach on ECDL (Enable/disable CMOS Differential Logic) logic [11] because it satisfies points 2 and 3 of the design procedure put forward in the previous subsection. ECDL gates have constant intrinsic delay regardless of the inputs and the function of the gate [12]. The ECDL is a clocked differential CMOS logic. An input clock that controls the two phases of the gate (at the disable phase the output nodes are precharged, and at the enable phase the logic function is evaluated) drives the ECDL gates. In a chain of ECDL gates the former generates the clock for the later gate, in a self-timed process. Using a regular structure satisfies design constraint 1.

The proposed low switching noise design strategy is applied to an unsigned 4 × 4 bits array multiplier [13]. The multiplier structure consists in a rectangular array of 3×4 cells (see Figure 9.5). Each row simultaneously computes a partial sum of two numbers of 4 bits. The result is propagated to the next row. The clock signal is also propagated from one row to the next in a self-timed process. There are two different cells in the structure: an external cell (used in the left and upper edges of the array) and an internal cell, both implemented using CMOS ECDL logic. The logic functions for the external (S_{ext} and C_{ext}) and internal (S_{int} and C_{int}) cells are as follows:

$$
\begin{aligned}
S_{ext} &= (a_{i+1} \cdot b_j) \oplus (a_i \cdot b_{j+1}) \oplus C_{ij} \\
S_{int} &= (a_i \cdot b_{j+1}) \oplus (S_{i+1j}) \oplus C_{ij} \\
C_{ext} &= (a_{i+1} \cdot b_j \cdot C_{ij}) + (a_i \cdot b_{j+1} \cdot C_{ij}) + (a_{i+1} \cdot b_j \cdot a_i \cdot b_{j+1}) \\
C_{int} &= (a_i \cdot b_{j+1} \cdot C_{ij}) + (S_{i+1j} \cdot C_{ij}) + (a_i \cdot b_{j+1} \cdot S_{i+1j})
\end{aligned}
\qquad \textit{Eq. (9.4)}
$$

where *a* and *b* are the bits of operands A and B, *i* and *j* are the multiplier array column and row indexes, respectively, and *C* and *S* are the preceding sum and carry bits. Each cell of the ECDL multiplier has a current limiter at Vdd and Gnd supply nodes that can be controlled by two voltage references: *VrefP* and *VrefN* (Figure 9.4.b). These current limiters are used to set the maximum switching current amplitude of the structure's single cells.

Another 4×4 bit array multiplier is designed using conventional static CMOS logic gates to implement the internal and external cell of the structure in order to compare the switching current waveforms obtained with the ECDL low switching noise multiplier and this static CMOS multiplier.

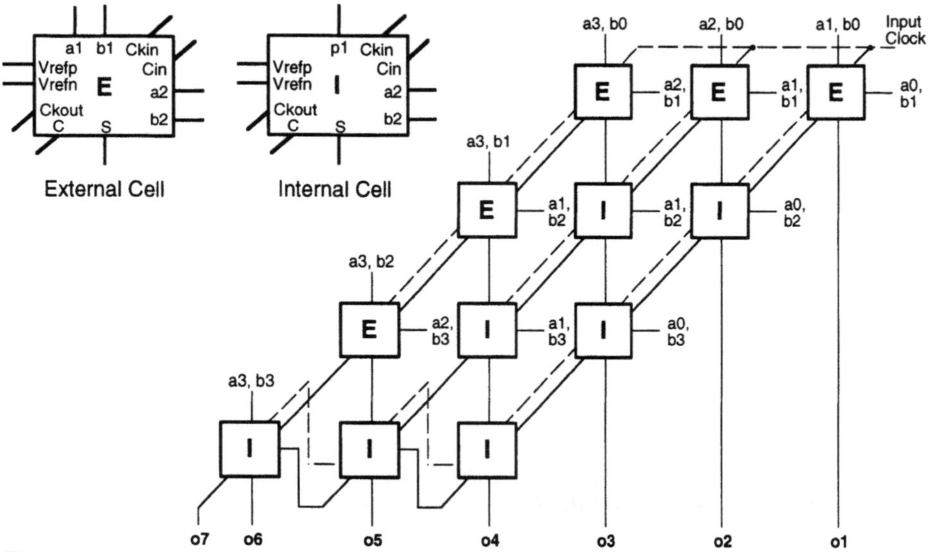

Figure 9.5

Structure of the 4x4 ECDL multiplier.

The simulation results in Figure 9.6 show the comparison between the switching current waveforms obtained for the two multipliers. The static CMOS multiplier has non-uniform switching current in each row and this generates an abrupt total switching current waveform. The switching current waveforms of the various rows in the ECDL multiplier are identical (except for the last row) and this produces a total switching current waveform very similar to the ideal trapezium shape.

The two multipliers were manufactured using a two metal, single poly 0.7μm CMOS process. The circuit also contains some additional logic (see Figure 9.7). A set of inverters is required to generate the complemented inputs for the ECDL multiplier.

The output of the two multipliers is multiplexed and a single 8-bit bus is available from the external outputs. A second multiplexer is used to measure the extra delay introduced by the first multiplexer and the input and output drivers. A photograph of the prototype manufactured is shown in Figure 9.8. The two multipliers have dedicated positive and ground terminals. The additional logic and the input/output drivers also have dedicated power supply terminals. The circuit was packaged using a 40 pin ceramic DIL and connected to the testing board with a socket. The outputs are monitored using 750 MHz and 1 GHz-bandwidth probes through low ground inductance probe tip adapters. The switching current waveforms for the ECDL and the static CMOS multiplier are measured off-chip using 1 GHz-bandwidth current probes.

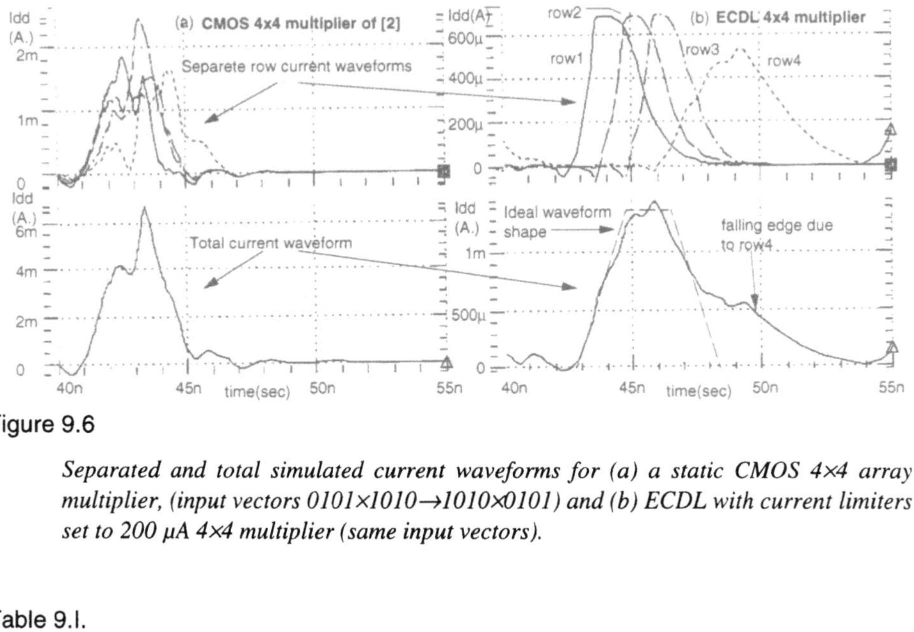

Figure 9.6

Separated and total simulated current waveforms for (a) a static CMOS 4×4 array multiplier, (input vectors 0101×1010→1010×0101) and (b) ECDL with current limiters set to 200 µA 4×4 multiplier (same input vectors).

Table 9.I.

Simulation results for three type of 4×4 bit multipliers. The power and noise results are obtained using a periodic input pattern: 0000×0000→ 1111×1111.

Multip. Type	Vdd peak noise (mV)	Gnd peak noise (mV)	Row delay (ns)	Total delay (ns)	Power@ max freq. (mW)	Power@ same freq. (mW)	# transistors (pmos+nmos)
ECDL+Limit.	25	20	1.35	7.29	2.684	2.49	99 + 427=526
ECDL	150	100	0.85	4.01	2.971	4.76	87 + 415=502
static CMOS	408	500	-	4.73	5.27	12.90	297 + 297=594

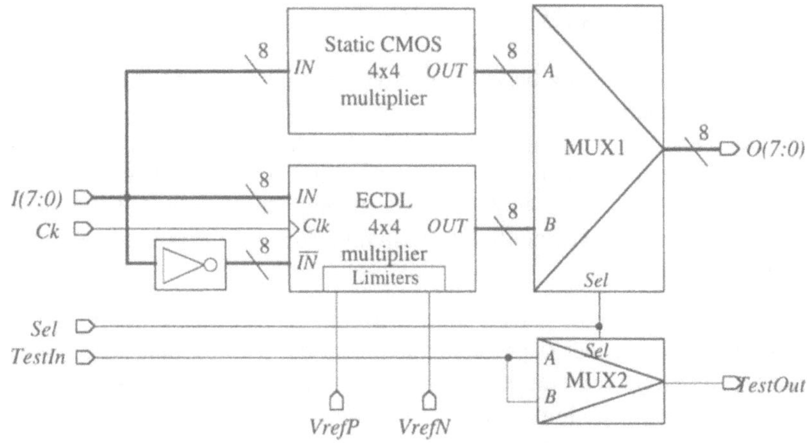

Figure 9.7
Structure of the prototype manufactured.

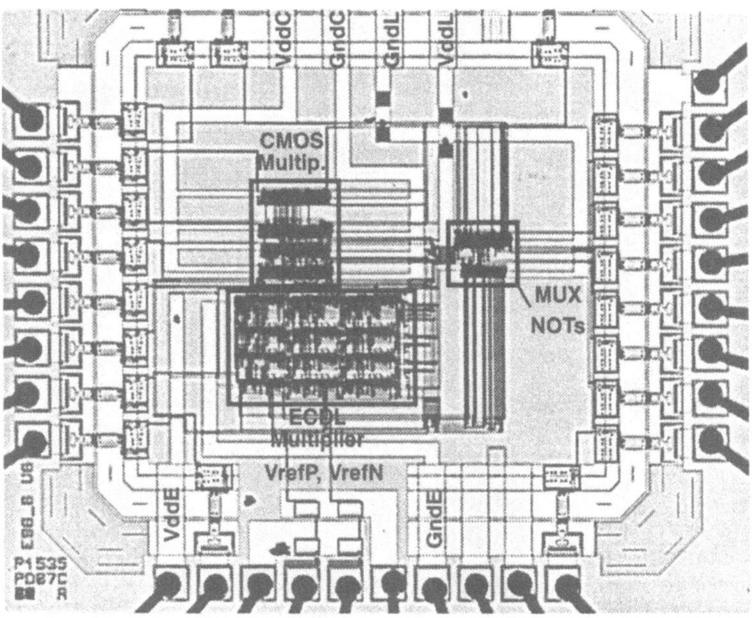

Figure 9.8

Photograph of the prototype manufactured.

9.3.3 Measurement results

Figure 9.9 shows a comparison between the switching currents of the two multipliers. The static CMOS multiplier has a switching current pulse of 8.1 mA maximum amplitude and has many peaks and valleys. The switching current pulse of the ECDL multiplier is much more regular. The maximum slopes for the CMOS and the ECDL multipliers are 1.37 A/μs and 0.46 A/μs, respectively. This means a 66.4% switching noise reduction for the ECDL multiplier, assuming the same power supply effective inductance in both cases.

Noise reduction is achieved by the smooth current waveform obtained in the ECDL multipliers. It is not a perfect trapezoidal waveform but the constant delay and shape of the individual current waveforms and the regular structure of the circuit help these to overlap properly to minimize the Delta-I noise generated. The suppression of glitching also helps to reduce switching noise.

The delay penalty due to the limited current constraint and the reduction in noise is a trade-off. As the current limit is raised the delay decreases, approaching the unlimited ECDL multiplier delay, but the noise increases (see Table 9.II).

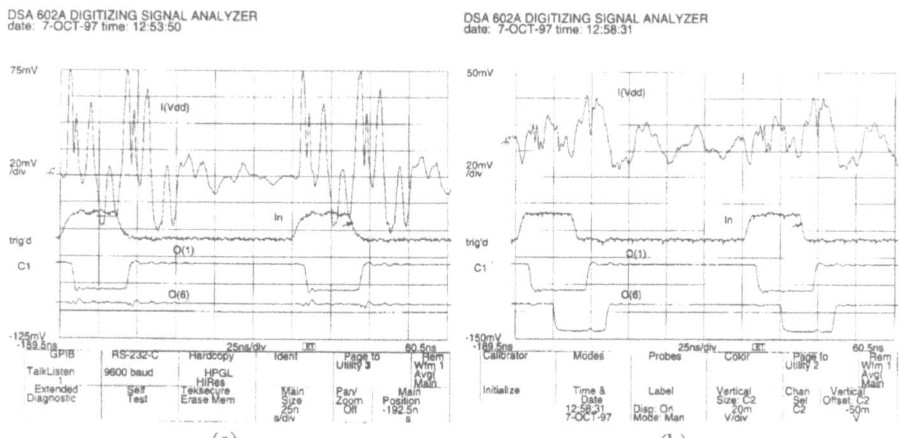

(a) (b)

Figure 9.9

Switching current waveforms: (a) Static CMOS multiplier with input pattern 1111x1111 → 1101x0110. (b) ECDL multiplier with input pattern 1101x0110, VrefP = 0 V and VrefN = 5 V. Current scale is 1.33 mA/div.

Table 9.II.

Effects of current limiters on the ECDL multiplier enable delay and switching current. The input pattern for these measurements is 1111×1111.

Limited current per cell	Enable delay from *Ckin* ⇓ to *O(6)*	Peak to peak switching current value
Maximum (370 μA)	18.9 ns	4.6 mA
200 μA	23.0 ns	3.3 mA
150 μA	28.6 ns	2.7 mA
100 μA	45.1 ns	3.0 mA

9.3.4 Conclusions

A straightforward design methodology for implementing low noise combinational circuits has been put forward. Regular structures make better use of this approach, though any logic structure could be well-suited by introducing redundant blocks. The delay penalty can be avoided by taking advantage of the self-timed behavior of the ECDL logic. In the simulation example, the maximum operating speed of the static CMOS multiplier is limited by its total delay (4.73 ns). In the ECDL+Limiters approach the maximum speed is limited by the two-row delay (2.7 ns) instead of the total delay, because once the second row of cells has computed and propagated the result to the next row the first row can compute the next input vector. By adding delay cells to the final outputs generated in each row, input vectors can be introduced, as in wave pipelining circuits, thus achieving a higher throughput.

This technique allows a considerable reduction of dI/dt switching noise to be obtained using a simple design procedure. It has several advantages over other techniques for reducing switching noise such as increasing the number of power pins and pads and the inclusion of on-chip decoupling capacitors that increase packaging costs, or the use of FSCL logic, entailing constant power consumption. The proposed design procedure can be applied to many digital circuits but is especially suitable for regular arithmetic or logic circuits and iterative networks. Indeed, by adding dummy delay cells any kind of circuit structure can be regularized, at the expense of increased area and power consumption. The ECDL self-timed operation helps to eliminate intermediate latches (that may switch simultaneously) from the pipeline, thus reducing the dI/dt switching noise even further.

9.4 Low switching noise power supply techniques

As mentioned above, the switching noise dI/dt depends on two factors. One is the generation factor that comes from circuit activity and depends on the waveform of the switching currents required of the supply system. This has already been dealt with in previous sections. The second factor is due to the inductive nature of the power supply system that converts the dI/dt into voltage noise. There are several switching noise reduction techniques based on minimizing the inductive term of the electronic systems' supply network. These techniques are used to assist proper design of supply voltage distribution, including the connectors and the structure of the printed circuit board planes; the value and placement of decoupling capacitors; package design; the design of the terminals used for the IC supply; and finally the distribution of the supply within the IC itself.

The distribution network of supply voltage through an electronic system can be seen as a hierarchic structure with several levels. The highest is at system level, the next being board level, then package level, module level and finally the distribution within the chip [14]. A simplified model of the power supply network with discreet elements can be used to study the main characteristics of switching noise. Switching noise reduction has to be addressed on all levels of this power supply hierarchy. In this section the overall structure of the supply network and the use of decoupling capacitors will be analyzed. Packaging techniques for reducing switching noise will also be considered. Next we will approach the problem of switching noise and supply distribution in the interior of monolithic CMOS ICs, analyzing the switching noise reduction techniques in the literature and presenting a new proposal complete with experimental results.

9.4.1 Power supply distribution and decoupling capacitors

The power supply distribution of an electronic system starts with the power supply source (PS) and the subsequent regulator (DC-DC converter). From this point the supply voltage is distributed (through cables, or supply buses in the case of a rack environment), up to the connectors of whatever different boards form the system. Each board can carry its own local regulator. In this case, the supply system to analyze for switching noise questions would begin after such a local regulator. The main mission of the supply system is to provide the circuits with stable voltages regardless of their consumption (static and dynamic). This stability is based on the fact that the supply net may have a low, constant impedance for the whole range of relevant frequencies. The aim of inserting decoupling capacitors is to compensate for the inductive nature of the conductors from different frequencies. The same structure found at board level can be applied, scaled, on MCM level.

Regulator output impedance can be taken to be an ideal voltage source with a small series output resistance. The distribution cables at system level have one inductive and one resistive component. Each conductor can be modeled as an inductor in series with a resistance. The value of the resistance will drop as the length is reduced and increase when the transverse section is reduced. The inductance also drops when the length is reduced. Its dependency on the cross-section is more complex and depends on the type of conductor. In a cylindrical section cable surrounded by an insulator or parallel to a conducting plane the inductance drops as the radius of the conductor is increased and rises as the radius of the insulation or the distance to the plane, respectively, are reduced. If the conductor is a metallic track with a rectangular section over a conducting plane the inductance drops with the width and increases when the distance from the conducting plane is reduced. There is a compromise when reducing the transverse section of the conductors, that makes both the resistance rise and the inductance drop (here reference is always to the self-inductance term of the inductor). The decoupling capacitors behave capacitively only at low frequency. For high frequencies above their resonance frequency they behave inductively [15][16]. Each level of the hierarchy can thus be represented as a conductor section that ends in a decoupling capacitor (or capacitors).

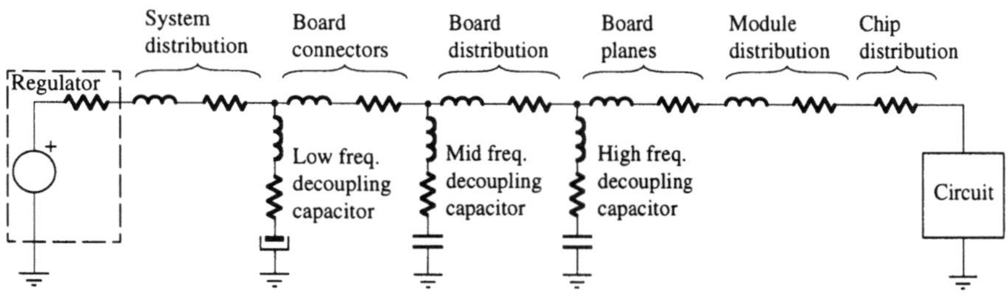

Figure 9.10

The hierarchy of power supply distribution.

In the design of supply distribution it is crucial to calculate the place [17] and the value of the decoupling capacitors properly. Figure 9.11 shows the impedance with respect to the frequency obtained after the power supply design process that is presented in [18]. The specifications for this system require an impedance below 18mΩ from DC to 35 MHz due to the fact that the circuit to be powered may present current consumption demands with variations of 0 to 10 A in 10 ns. Table 9.III gives the values of the different elements according to the general structure of Figure 9.10. In this example we remain at printed circuit board level. The voltage regulator is located next to the board, so that inductance due to the power supply distribution at system level is eliminated. Another particularity of this example is that the output capacity of the regulator (C_O) itself is used as the low-frequency decoupling capacitor.

Table 9.III.

Values of the components of the supply distribution system for the system presented in [18].

Regulator	Regulator Decoupling cap.	Regulator-PCB connectors	PCB lines	PCB decoupling cap.
$R_O = 15\ m\Omega$	$R_{CO} = 10.4\ m\Omega$		$R_S = 3.6\ m\Omega$	$R_{CB} = 4\ m\Omega$
	$L_{CO} = 1.25\ nH$	$L_C = 1.0\ nH$	$L_B = 0.5\ nH$	$L_{CB} = 63\ pH$
	$C_O = 6000\ \mu F$			$C_B = 30\ \mu F$

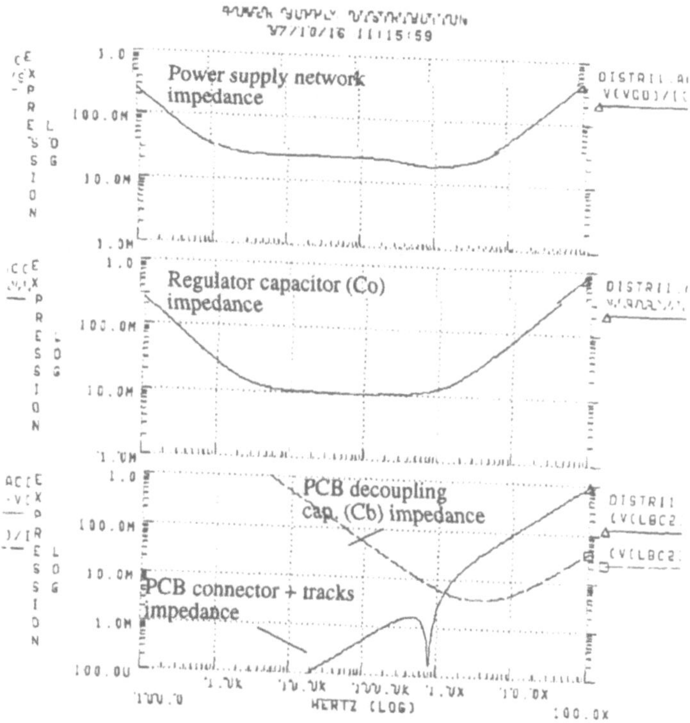

Figure 9.11

Power supply network impedance for the values given in Table 9.III.

The effect of the parasitic items of the decoupling capacitors can be clearly appreciated in Figure 9.11. The inductance of the two capacitors (C_o and C_B) is shown and it will be appreciated how these behave inductively above a given frequency (roughly 1 MHz for C_o and 10 MHz for C_B). A flat pattern for power supply impedance is achieved from 1 kHz to almost 10 MHz. To increase the band in which the supply network impedance is kept constant and under a given low value it would be necessary to add another bank of decoupling capacitors for the highest frequency, with better performance as regards high frequency behavior. In any case, these capacitors would be located inside the module or the package to compensate for the inductive behavior of the capacitor C_B and the inductance of the package terminals at the same time.

The design of supply distribution inside a module or an integrated circuit or chip entails greater complexity due to the smaller size of the package. The common substrate can be seen as a basically resistive mesh [19][20]. In this case the problem is resistive noise, proportional to the current pulses and not to its derivative. Low noise design is based on detecting points of the module where the supply voltage drop due to resistive or inductive noise (dI/dt) is important, and then calculating the optimum placement of the internal decoupling capacitors. In all these systems, however, the dominant inductance is that of the package, which we will analyze in the following subsection.

9.4.2 Packaging and switching noise

The design and selection of the package is vital for reducing switching noise. At PCB level there is a good deal of design freedom so it is not difficult to obtain some good characteristics of the supply distribution system (high values of the decoupling capacitors, capacitors of different types and sizes, etc. can be used). The only limitation is the inductive behavior of these decoupling capacitors for high frequencies. However, inside the package the size of the decoupling capacitors is limited by the package's physical dimensions [21]. Due to the fact that switching noise is directly proportional to supply terminal inductance, minimizing this inductance is a high-priority in the design of ICs.

There are two alternatives for minimizing package inductance. First, several package terminals dedicated to supply voltage can be used. In a first approximation, the final value of the inductance will be the inductance of a single terminal of the package divided by the number of supply terminals. This first alternative raises the package cost as the ratio between supply pins and I/O pins grows higher (i.e., the microprocessor of [22] requires 187 supply terminals out of a total of 520). There are several methods published in the relevant literature for obtaining the optimum number of terminals used for the power supply and selecting these from all the terminals of the package [23][24]. Lead frame packages, such as the DIP, SOIC or QSOP have extremely varied pin self inductance values depending on whether this is in the central part of the package or in the corners. If increasing the number of power supply pins is not sufficient or cost-effective the type of package must be replaced by another with a smaller terminal inductance. Table 9.IV shows different packages and other advanced interconnection methods between the chip and the board, such as the solder balls used in flip chip techniques or the direct wiring of the chip on the board. The data was taken from several sources: [14][25][26][27].

The table shows how the surface mount packages (SMT, SOIC or QSOP) display less inductance that those of through-the-hole mount packages (DIP and PGA). Advanced packages such as those of BGA type have reduced inductance values of a few nH [27]. To reduce inductance still further flip chip techniques or direct wiring on the board have to be used, thus obtaining an inductance per terminal of between 0.1 nH and 1 nH. The exact cost of these alternatives is not readily available but is much higher than previous package techniques since they need machinery for assembling the chip on the board and, in the case of flip chips, very sophisticated package and material manufacturing technologies. This very important trade-off between the cost of the package and the minimum required inductance must be approached in the IC design process.

Table 9.IV.

Electrical parameters of the different packaging alternatives.

Package	Typical inductance	Typical capacity
DIP 68 plastic	35.0 nH	4.0 pF
DIP 68 ceramic	20.0 nH	7.0 pF
SMT 68	7.0 nH	2.0 pF
PGA 68	7.0 nH	2.0 pF
PGA 256	15.0 nH	5.0 pF
DIP 18 plastic	13.7 nH	
SOIC 18	8.5 nH	
QSOP 18	3.6 nH	
BGA	3 nH	
Wire bond	1.0 nH	1.0 pF
Solder ball	0.1 nH	0.5 pF

9.5 Reducing switching noise by using on-chip decoupling capacitors

Techniques for reducing switching noise after establishing package inductance focus on the design of the decoupling capacitors or other similar techniques within the package (on-chip). In this section the design of decoupling capacitors and their efficiency is analyzed first of all, on a simplified electrical representation of the IC and package. It is proved that the LC circuit formed by package inductance and decoupling capacitor produces oscillations that can be more harmful than switching noise itself [28]. The different alternatives proposed in the literature for solving these problems are summarized below. Finally an original proposal is presented as the contribution of this thesis work to minimizing switching noise through appropriate supply techniques.

Figure 9.12.a shows a simplified representation of a CMOS IC. Some of the logic gates switch while the rest keep their output fixed to '0' or to '1'. A decoupling capacitor (C_D), that will be considered ideal for the time being, is added. The two inductors with value L represent the inductance of the supply terminals. [14] shows how this schematic can be simplified by the simplest circuit of Figure 9.12.b. R_C represents the resistance in parallel of all the NMOS or PMOS transistors that do not commute and C_C is the non-commuted capacity, that can be calculated as $(1\text{-}\alpha)\cdot C_{CIR}/2$, where α represents the activity of the system and has a value between 0.05 and 0.3 [29][30], and C_{CIR} is the total capacity of the IC [31]. It can easily be found that the part that does not commute forms a RLC circuit with the following resonance frequency (w_o) and extenuation factor (ξ):

$$w_o = \frac{1}{\sqrt{\underbrace{2L}_{L_T}\underbrace{(C_D + C_C)}_{C_T}}} \qquad \xi = \frac{R_C}{2}\sqrt{\frac{C_T}{L_T}} \qquad \textit{Eq. (9.5)}$$

The part of the IC that commutes can be represented by a current source that demands a pulse of a given shape. In the analysis that is performed below a triangular current pulse of amplitude I_{max} and rising time t_r is used. Figure 9.12.d shows the waveform of the internal Gnd terminal when a switching occurs. The waveform of the internal Vdd terminal is vertically symmetrical to the previous one and centered on the constant Vdd voltage value. The subtraction of these two waveforms — the switching noise at the terminals of R_T and C_T — consists of a first minimum due to simultaneous switching noise (SSN) and an oscillation that is generated next with the fundamental frequency of Eq. (9.5).

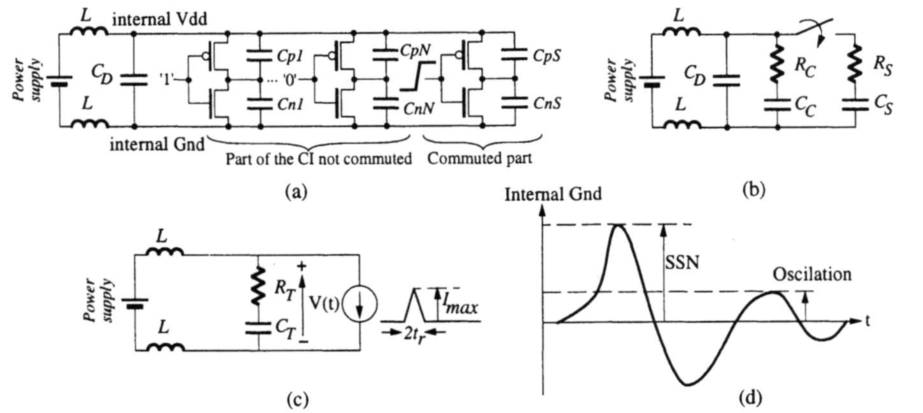

Figure 9.12

(a) Model of a packaged IC with one part that does not commute and another that commutes. (b) Simplified electrical model of the previous circuit. (c) RLC circuit formed by the part of the IC that does not commute and the package inductances. (d) Waveforms at the internal supply terminals when a switching event occurs in the IC due to simultaneous switching noise (SSN) and to RLC circuit oscillation.

The important part of this analysis is determining the peak values of both minimals (the one due to SSN and the one for oscillation) and studying their maximum levels. In a first approximation to the problem we will suppose that $R_C = 0$ and afterwards study the effect of the damping introduced in the oscillation by R_C. The transient analysis of the circuit in Figure 9.12.c can be split into three parts depending on the switching current pulse.

1) $0 < t \leq t_r$: The value of the capacitor voltage is as shown in Equation (9.6). Also the value of the minimal due to SSN and the maximum for this value is indicated:

$$V_{C_T}(t) = V_{DD} - \frac{I_{max} L_T}{t_r}\left(1 - \cos\left(\frac{t}{\sqrt{L_T C_T}}\right)\right)$$

$$\min\left(V_{C_T}(t)\right) = V_{DD} - \underbrace{\frac{I_{max} L_T}{t_r}\left(1 - \cos\left(\frac{t_r}{\sqrt{L_T C_T}}\right)\right)}_{SSN} \qquad Eq.\ (9.6)$$

$$SSN_{\max} = \frac{2 I_{max} L_T}{t_r} \qquad t_r = k\pi\sqrt{L_T C_T},\ k = 1,3,5,\ldots$$

2) $t_r < t \le 2t_r$: This part is not relevant, since it contains only the falling edge of the circuit's switching current pulse.

$$V_{C_T}(t) = V_{DD} + \frac{I_{\max} L}{t_r}\left(1 - \left(2 - \cos(t_r w_o)\right)\cos\left((t - t_r)w_o\right) - \sin(t_r w_o)\sin\left((t + t_r)w_o\right)\right)$$

Eq. (9.7)

3) $t > t_r$: From this time there is harmonic oscillation whose extent can be calculated from the energy that is left stored in the inductances (L_T) and in the capacitors (C_T) after the current pulse that represents the forced excitement of the system. The extent of this oscillation and its maximum level is given by the following equation:

$$V_{C_T}\Big|_{pico} = \frac{I_{max} L_T}{t_r} 2\left(1 - \cos\left(\frac{t_r}{\sqrt{L_T C_T}}\right)\right) \qquad \textit{Eq. (9.8)}$$

$$\max\left(V_{C_T}\Big|_{pico}\right) = \frac{4 I_{max} L_T}{t_r}$$

where the maximum level for the peak value of V_{CT} is given for the same conditions as in Eq. (9.6). This maximum level for the harmonic oscillation is double that of the switching noise. From these equations a criterion for calculating the minimal decoupling capacitor can be established. The worst case is given for a value of $C_T^* = (t_r/\pi)/L_T$. If $C_T \gg C_T^*$ Equations (9.6) and (9.8) can be approximated by:

$$SSN \propto I_{max}\sqrt{\frac{L_T}{C_T}} \qquad V_{C_T}\Big|_{pico} \propto 2I_{max}\sqrt{\frac{L_T}{C_T}} \qquad \textit{Eq. (9.9)}$$

Figure 9.13 illustrates this situation, where the currents across each of the items and the voltage noise on the capacitor C_T for the circuit in Figure 9.12.c with $R_T = 0$ have been drawn.

The existence of this oscillation has very harmful effects if w_o is an entire multiple of the system's clock frequency, since it can produce noise accumulation when the SSN of current switching is added to the previous switching oscillations [31][32]. The oscillation can provoke failures in other chips or circuits connected to the noisy circuit through the output drivers, apart from affecting the circuits integrated in the same chip, as shown in [32]. To prevent this, such oscillations have to be attenuated. The resistance available in the circuit to attenuate these oscillations, R_T, tends to be small, in the range of a few ohms. Figure 9.14 shows simulations performed with different values of R_T for C_T values above and below C_T^* for that circuit.

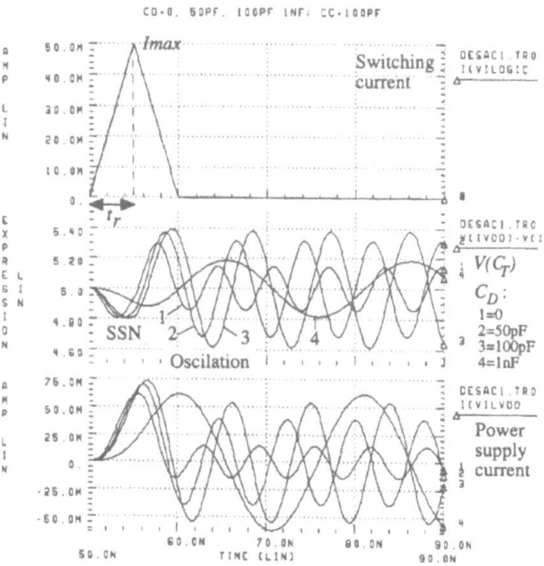

Figure 9.13

Simultaneous switching noise (SSN) and oscillation for the circuit from Figure 12.b with I_{max}=50 mA, t_r=5 n, L_T=10 nH, C_C=100 pF, C_T^*=253.3 pF and R_C=0 and different decoupling capacitor (C_D) values.

9.5.1 On-chip decoupling capacitors

The unswitched internal capacitance in an IC acts as a decoupling capacitance [33]. Figure 9.15 shows the results of an example: a 4x4 bit multiplier implemented with conventional static CMOS logic. The layout for this circuit is drawn, including the input and output pads. Next the electrical circuit, including the parasitic capacitors due to the interconnections, was extracted from the layout. The circuit obtained was connected to the power supply in the simulator netlist through a resistance (substantially larger than resistance R_T of the model used in the previous subsection). Measurement of the internal unswitched capacitance is performed as follows. A given input vector is established and then the voltage of the power supply is suddenly cut from 5 to 0 volts. The supply node-discharging time constant is measured. This time is used along with the resistance of known value to deduce the capacity. This measurement was obtained by simulation for all the possible input combinations (256 different vectors) and the results are shown in Figure 9.15.b, in the form of a histogram. For this circuit the mean capacity found is 72.2 pF and its standard deviation is ±3.35 pF.

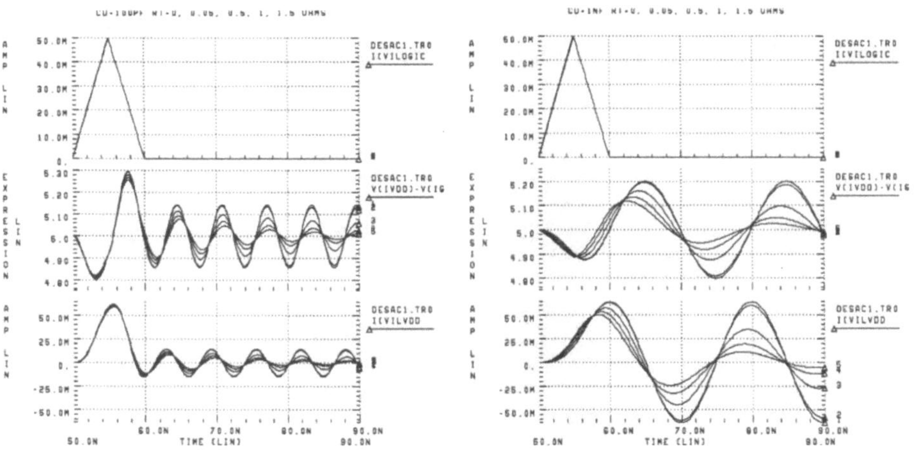

(a) (b)

Figure 9.14

Waveforms of the switching current, the current of the power supply and the internal supply voltage Vdd-Gnd of the circuit in Figure 12 for two C_T values: (a) 100 pF and (b) 1 nF and R_T values rising from 0 to 1.5Ω..

The unswitched IC's capacitance is normally insufficient to act as a decoupling capacitor alone. To reduce the levels of switching noise, one customary practice is to integrate a capacitor within the [34][35][36][37]. The capacitor can be included within the cavity of the package, or over or under the silicon chip, and be bond wired between the supply terminals [36][38]. In multichip modules the on-chip capacitor can be another of the chips and be connected to the common substrate, or wire bonded, or soldered using solder balls [19]. These forms of adding the decoupling capacitor have high frequency inductive behavior as a limitation. The problem is more serious in bond wired alternatives than in those that use solder balls. For very high frequency applications the capacitor must be integrated in the same to minimize the length of the interconnections between the decoupling capacitors and the internal supply terminals or planes of the IC [19][36]. The semiconductor substrate behaves capacitively and is thus a good decoupling capacitor up to frequencies of 500 MHz [39]. The efficiencies reported in the literature on the reduction of SSN within the package with and without internal decoupling capacitor range from 30% to 70% [34][36].

However, on-chip decoupling capacitors are not useful for reducing the switching noise generated by the output drivers [14]. The current loop is closed by only one of the two supply terminals when external loads have to be switched; the positive for low to high level transitions and the negative for the complementary transitions.

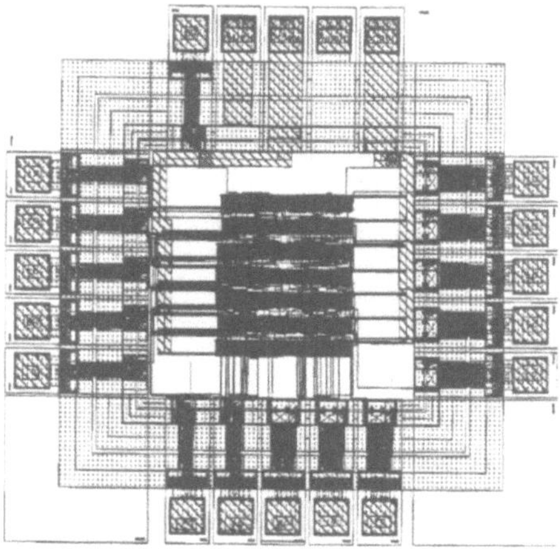

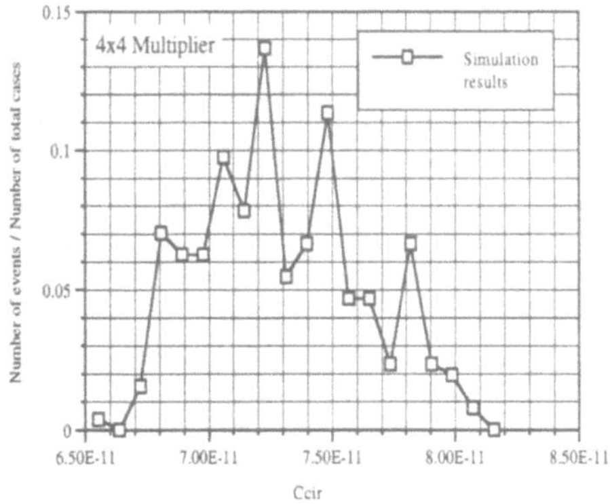

(a) (b)

Figure 9.15

(a) Layout generated by the CAD tool for the 4x4 CMOS multiplier made with 1.0μm CMOS technology. (b) Histogram of the circuit's unswitched capacitance for all possible input combinations.

9.5.2 Solutions for the problem of the oscillations produced by on-chip decoupling capacitors.

As has already been seen, the inductances of the package and the on-chip decoupling capacitor together with the capacity of the non-commuted part of the IC form an RLC circuit that generates a damped oscillation each time a switching event occurs.

The resistances that can be found due to the intrinsic elements of the IC (resistance of the transistors that do not commute, package wiring, etc.) are not more than a few ohms. If this resistance is under what is required to assure a given damping it will be necessary to add supplementary resistances in the supply lines to increase the damping factor (ξ). This added resistance introduces a new type of noise, the resistive noise proportional to *I(t)*, the switching current. In [31] different alternatives for introducing this additional damping are analyzed, including the use of active resistances serially with the decoupling capacitor. In spite of the reported reduction of dI/dt switching noise and the reduction of RLC oscillation, the authors themselves admit that their proposal causes an increase in resistive noise.

A second alternative recently appearing in the literature [40] proposes using a supplementary RLC circuit connected in parallel with the on-chip decoupling capacitor of the RLC circuit of the IC + package together so that this compensates for the original resonance frequency. To work properly there has to be a relationship between the supplementary inductance and capacity and the original RLC elements of the IC. The main problem is that the contribution of the non-commuted capacity to the capacitive term of the original circuit is variable, as was seen in the previous subsection with the conventional CMOS 4x4 multiplier. If the on-chip decoupling capacitor is not much greater than the IC's internal capacity, the latter's variability due to the different logic inputs of the circuit can overcome the compensation introduced by the supplementary circuit. In MCM circuits with several on-chip decoupling capacitors located at different points of the package there are not just one, but several resonance frequencies of the RLC circuit. It is not possible to compensate these properly with a single supplementary RLC circuit. This technique is however very useful for monolithic circuits with an on-chip decoupling capacitor much greater than the IC's internal capacity.

9.6 Low noise power supply technique using an on-chip tank capacitor: TCMOS

In the previous sections of this chapter several techniques for reducing switching noise have been examined. Increasing the number of power supply pins reduces parasitic inductance. On-chip decoupling capacitors are also frequently used. The former technique involves a sharp increase in the packaging cost (in current microprocessors nearly half the pins are used for power supply connections). The latter entails the drawback of the oscillations produced by the LRC circuit formed by the package inductors, the decoupling capacitor and the logic circuits. Our proposal, the TankCMOS technique, is based on placing intermediate circuitry between the Vdd internal pin and Vdd node of the digital circuitry (see Figure 9.19.16). It uses an on-chip capacitor that is effectively insulated from the internal pins of the package's power supply terminals. Transient switching current pulses are prevented from flowing across the inductive package parasitics (Figure 9.16.b). The problem of resonance is also solved. The authors first presented this technique in 1995 [41]. Recently the TCMOS principle has been applied to noise suppression in Gigabit DRAMS [42]. This section presents an improved version of the TCMOS circuit and a practical implementation and measurements.

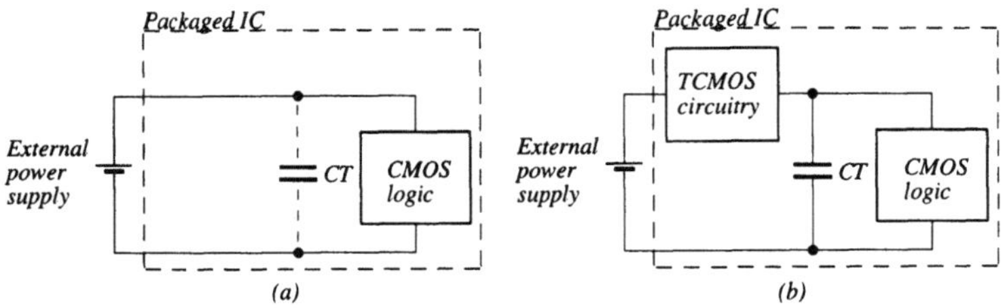

Figure 9.16

(a)Traditional power supply approach with the possibility of including an on-chip decoupling capacitor. (b) TCMOS technique block diagram

9.6.1 The TankCMOS (TCMOS) approach

The *Tank CMOS* (TCMOS) low noise power supply circuit operates as illustrated in Figure 9.17. The on-chip decoupling capacitor (tank capacitor or CT) directly supplies the logic. However, it is not directly connected to the external power supply. A switched current mirror is used to charge the capacitor using a constant current when the voltage in the capacitor drops below a specified threshold voltage (Vmin). Then the voltage of the capacitor increases until it reaches the upper limit (Vmax). At this time the current mirror is turned *off* and the capacitor is isolated from the external power supply. When the mirror is *on* it behaves like a constant current source, preventing current transients from flowing through the external power supply and the package power supply pins and wire bonds (see the measured waveforms in Figure 9.19). A Schmitt trigger that senses the voltage across the tank capacitor controls the mirror switching. This Schmitt trigger is made up of three inverters with different switching thresholds and an RS latch. The voltage is sensed across a voltage divider (Figure 9.17.b). The different threshold configurations are selected through two digital control inputs.

The TCMOS circuit can be designed always to be in *on* when the logic section is working. The current flowing across the package inductances is then constant and switching noise generated by the internal logic circuitry is eliminated.

9.6.2 TCMOS circuit and functionality

Figure 9.20 shows the layout of the manufactured circuit using a 1.0μm ES2 CMOS technology. The tank capacitor is a SMD capacitor adhered to the cavity with non-conductive epoxy and is wire-bonded to the CMOS chip. The circuit contains a selection of combinational and sequential logic blocks. The logic is surrounded by a p+ diffusion guard ring. The terminals of the capacitor and the power supply nodes of the TCMOS circuitry (the mirror and the Schmitt trigger) are doubled in order to use one of the pins to supply and the second to measure the voltage waveforms at the on-chip nodes. The input and output drivers of the logic circuits are powered separately from the rest of the circuit.

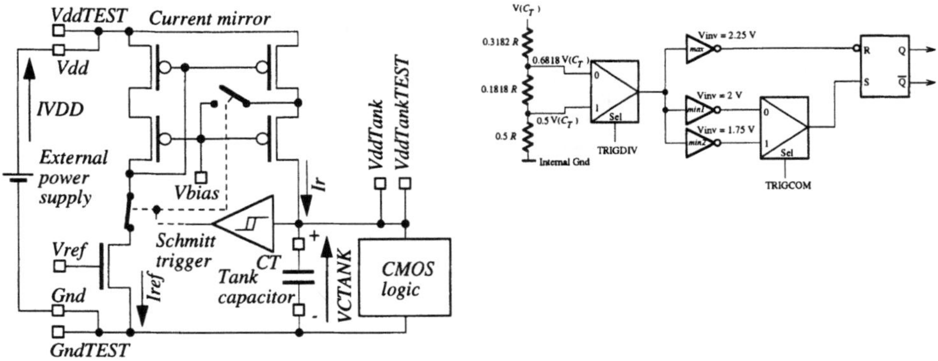

(a) (b)

Figure 9.17

TCMOS power supply circuit structure (a) and Schmitt trigger structure (b)

The relatively large area occupied by the current mirror of the general purpose prototype made can be reduced a great deal in an application-specific implementation by using a simple current mirror structure and reducing the output voltage constraints.

The different alternatives for power supplies to the digital logic in the prototypes are shown in Figure 9.18. In case A the TCMOS technique is used to supply the logic by connecting the external voltage supply to pins Vdd and Gnd. On-chip voltage supply noise is measured at the VddTEST pin and the logic supply voltage at the VddTankTEST pin. The logic is supplied by the tank capacitor (CT) placed on-chip between the VddTank and Gnd nodes. In case B the external voltage supply is connected to VddTank and Gnd pins and on-chip voltage supply noise is measured at the VddTankTEST pin. There is the same on-chip capacitor (CT) between VddTank and Gnd nodes as in the previous case. Case C is similar to case B, except for capacitor CT that is placed off-chip, on the measuring board. Case D is the same as case C without the off-chip decoupling capacitor. In cases B, C and D the TCMOS circuitry (current mirror and Schmitt trigger) is not powered and pin Vdd is floating. In case A (TCMOS technique) the charging current when the mirror is *on* can be set between 0 and 5 mA with an external voltage reference. There are four different limits [Vmax - Vmin] for the Schmitt trigger, and thus the following logic supply voltage swings: [3.5 V - 4.5 V], [4.0 V - 4.5 V], [2.6 V - 3.3 V] and [3.0 V - 3.3 V]. Twenty packaged samples of the same IC were made, five with an on-chip tank capacitor of

CT = 0.33 nF, five with CT = 4.7 nF, five with CT = 47 nF and five with no capacitor. Cases A and B are measured using samples of the first three sets of samples and cases C and D with the last set of samples.

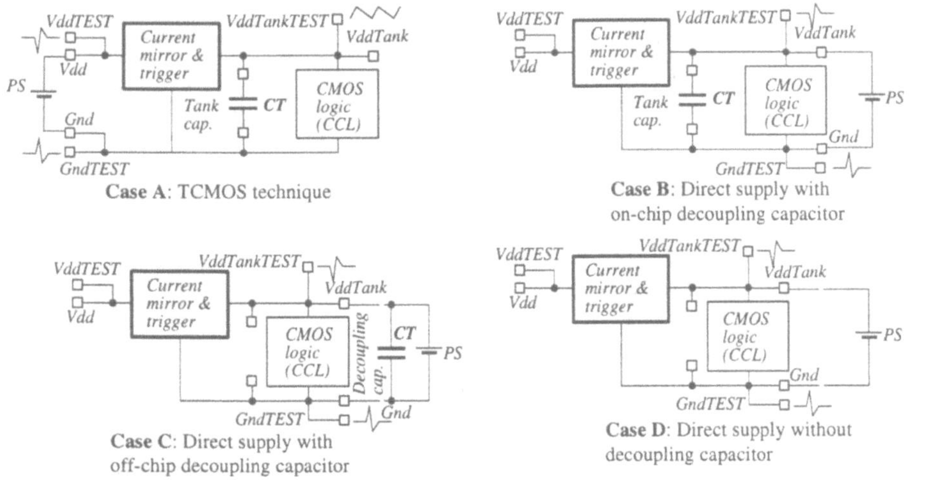

Figure 9.18

Possible power supply configurations available in the prototype.

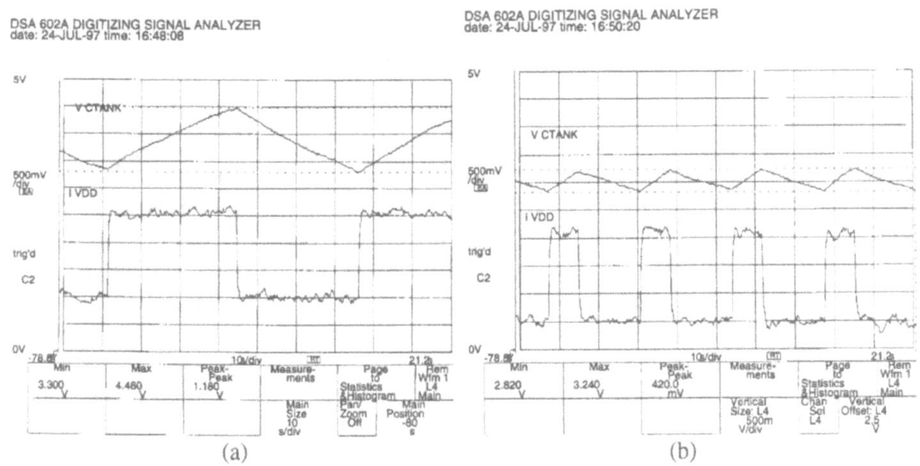

Figure 9.19

Measured logic circuitry supply voltage for the TCMOS power supply circuit CT = 47 nF. Scale for the external supply current waveforms (IVDD): 2 mA/div. (a) Supply margin from 3.5 V to 4.5 V; (b) from 2.9 V to 3.3 V.

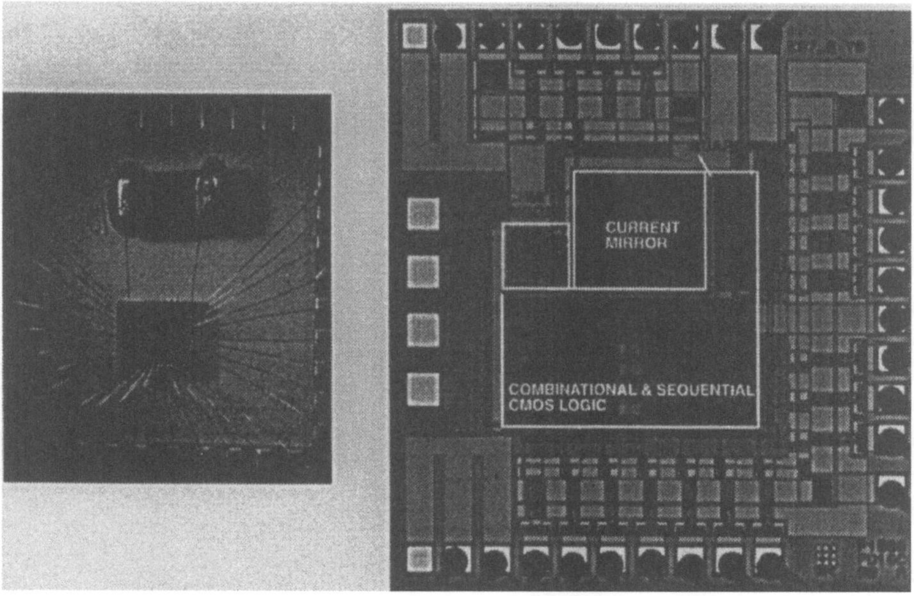

a) *b)*

Figure 9.20

(a) Photograph of one of the prototypes with tank capacitor in the cavity in the package. (b) Photograph of the ASIC in which the different parts of the circuit are indicated.

9.6.3 Results and discussion

Figure 9.21 shows the measured logic supply voltage swing and the external supply current of the TCMOS. In this case the selected voltage swings are [3.5 V - 4.5 V] and [2.9 V - 3.3 V]. The recharging and discharging times are set, choosing the recharging current and tank capacitor values, and these also depend on the selected voltage swing and the power consumption of the logic plus the power consumption overhead introduced by the TCMOS circuitry [41]. Figure 9.21 shows the noise waveforms measured at the on-chip power supply node in cases A (with a tank capacitor CT = 47 nF) and D, where the switching noise reduction achieved by using the TCMOS circuit is shown. The TCMOS circuit also gives higher noise reduction than the level obtained with an on-chip decoupling capacitor (case B) as shown in Table 9.V. Noise, measured as the maximum negative peak amplitude at on-chip V_{dd} node, is reduced by 18.82% by using a 47 nF on-chip decoupling capacitor. Noise is reduced by 70.64% with the proposed TCMOS circuit using the same on-chip tank capacitor value and oscillation problems at the on-chip power buses are also avoided. Noise reduction is slightly lower with smaller tank capacitors, as Table 9.V shows. Similar simulation

simulation results are reported in [42], which also vouches for the advantage of the power supply pincount reduction obtained by using the TCMOS technique.

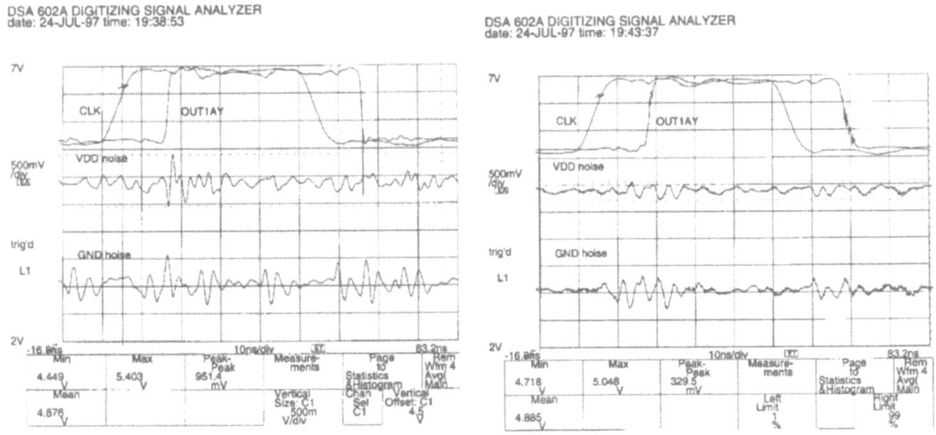

(a) (b)

Figure 9.21

Measurements of switching noise at the internal supply terminals. (a) Logic circuit directly supplied without decoupling capacitor (case D). (b) Logic circuit supplied with TCMOS technique with C_T = 47 nF and I_r = 4 mA.

Table 9.V.

Measured negative voltage noise peak at the on-chip Vdd node for different capacitor values in the four supply cases given in Figure 9.18.

Capacitor:	47 nF	4.7 nF	0.33 nF
Case A (TCMOS)	145 mV	156 mV	164 mV
Case B	401 mV	423 mV	465 mV
Case C	441 mV	—	—
Case D	494 mV		

Table 9.VI presents the noise measured at the guard ring that surrounds the logic. This guard ring can either be connected to the ground plane of the measurement board through a dedicated pin or be left floating The substrate noise collected by the guard ring can be measured by a second pin. The noise is reduced by 26.15% if the ring is biased with the Gnd voltage in case D (no decoupling capacitor and logic directly supplied by the external power supply). However, if the TCMOS technique is used to supply the logic the noise is reduced by 35.89% even with the guard ring floating. The TCMOS also gives better substrate noise reduction than the on-chip decoupling capacitor.

Table 9.VI.

Measured positive voltage noise peak at the substrate. In cases A and B the on-chip capacitor is 47 nF.

	Guard ring grounded	Guard ring floating
Case A (TCMOS)	180 mV	250 mV
Case B	223 mV	310 mV
Case D	288 mV	390 mV

The proposed TCMOS circuit effectively reduces both switching noise and substrate noise due to the digital logic switching. Noise suppression is obtained without increasing the number of power pins. Resonances due to the package and on-chip capacitor are also avoided. The improved current mirror structure allows higher internal supply voltages closer to the external power supply voltage. It involves the drawback of a time-varying logic supply voltage which can, if required, be solved by inserting a regulated serial voltage down converter (VDC), as done in [42].

The tank capacitor placed in the cavity of the prototypes can easily be integrated in the same substrate as the logic up to values in the range of hundreds of nanofarads. The TCMOS can be adjusted to a wide range of IC sizes and power consumptions.

9.7 Summary

Several techniques and circuits for reducing switching noise have been analyzed in this chapter. In the first part switching noise reduction is achieved through controlling the time derivative of the total switching current. As an alternative to the method explained in Chapter 8, the aim is not to reduce the activity of the circuits but instead to implement them so that the switching current has some controllable characteristics, regardless of the circuit's function.

The first technique consists in using constant current consumption logic (CSL, ESCL, and FSCL). The power supply current is constant. The output voltage swing is also reduced, which favors speed and reduces coupling problems between signal lines (crosstalk). The main disadvantage of these logics is their high consumption.

The second design technique for minimizing noise aims to obtain a specific waveform of the switching current for a digital circuit similar to a trapezium. The results obtained by simulation as well as on a manufactured IC prototype demonstrate the usefulness of this technique. As opposed to the technique based on FSCL logic and similar ones, our proposal does not involve any increase in consumption, though it does increase the area.

The use of current limiters allows the levels of switching current generated to be controlled and reduced even further but is effective within certain suitable limits, as has been experimentally demonstrated, and entails the drawback of an increase in the delays.

The use of self-timed structures helps to do without registers and flip-flops, thus reducing the level of simultaneous activity synchronous with the clock. The speeds that can be reached with these techniques are higher than those of CMOS circuits since the propagation paths are perfectly controlled and the data gradually advances in a regular way through the structure in the form of waves. These techniques are named wave pipelines and also give lower power consumption since the possibility of spurious transitions or glitches is eliminated.

The second half of the chapter is devoted to techniques for reducing switching noise from the power supply design standpoint. Basic approaches consist in increasing the number of power pins or using advanced packages, which are costly. Switching noise can be reduced at system and board level by inserting decoupling capacitors. On-chip decoupling capacitors can also be used inside the package to reduce switching noise at module or IC level, but in this case the problem of self-resonance between the capacitor and the package appears. Special attention has to be paid to the value and position of these on-chip capacitors to avoid this problem. An alternative proposal to avoid the resonance problem when using on-chip capacitors is the TCMOS technique presented in detail in the last section of this chapter.

References

[1] David J. Allstot, Sayfe Kiaei, and Rajesh H. Zele, 'Analog Logic Techniques Steer Around the Noise', *IEEE Circuits & Devices*, September 1993, pp. 18–21.

[2] S. Kiaei, S.H. Chee, and D.J. Allstot, 'CMOS Source-Coupled Logic for Mixed-Mode VLSI', *Proc. IEEE Int. Symp. Circuits and Systems*, 1990, pp. 1608–1611.

[3] M. Maleki, and S. Kiaei, 'Enhancement Source-Coupled Logic for Mixed-mode VLSI Circuits', *IEEE Tr. on Circuits and Systems—II*, Vol. 39, No. 6, June 1992, pp. 399–401.

[4] D.J. Allstot and S. Kiaei, 'MOS Folded Source-Coupled Logic', U.S. Patent #5,149,992, Sept. 22, 1992.

[5] David J. Allstot, San-Hwa Chee, Sayfe Kiaei, and Manu Shrivastawa, 'Folded Source-Coupled Logic vs. CMOS Static Logic for Low-Noise Mixed-Signal ICs', *IEEE Tr. on Circuits and Systems—I*, Vol. 40, No. 9, September 1993, pp. 553–563.

[6] Sailesh R. Maskai, Sayfe Kiaedi, and David J. Allstot, 'Synthesis Techniques for CMOS Folded Source-Coupled Logic Circuits', *IEEE J. Solid-State Circuits*, Vol. 27, No. 8, August 1992, pp. 1157–1166.

[7] E. Alburquerque, J. Fernández, and M. Silva, 'NMOS Current-Balanced Logic', *Electronics Letters*, Vol. 32, No. 11, May 1996, pp. 997–998.

[8] Marc J. Loinaz, and Bruce A. Wooley, 'A BiCMOS Time Interval Digitizer based on Fully-Differential, Current-Steering Circuits', *IEEE J. Solid-State Circuits*, Vol. 29, No. 6, June 1994, pp. 707–714.

[9] Jose Luis González y Antonio Rubio, 'Circuitos Digitales CMOS de Bajo Nivel de Ruido dI/dt Basados en Lógicas Diferenciales y Limitadores de Corriente', *Seminario Anual de Automática y Electrónica Industrial*, Zaragoza, 11–13 de septiembre de 1996, pp. 246–251.

[10] Jose Luis González, and Antonio Rubio, 'Low Delta-I Noise CMOS Circuits Based on Differential Logic and Current Limiters', Accepted at *IEEE Tr. on Circuits and Systems Part I.*

[11] Shien-Lien L. Lu, 'Implementation of Iterative Networks with CMOS Differential Logic', *IEEE J. Solid-State Circuits*, Vol. 23, No. 4, 1998, pp. 1013–1017.

[12] Shien-Lien L. Lu, and Milos D. Ercegovac, 'Evaluation of Two-Summand Adders Implemented in ECDL CMOS Differential Logic', *IEEE J. Solid-State Circuits*, Vol. 26, No. 8, August 1991, pp. 1152–1160.

[13] Joseph J.F. Cavanagh, *Digital Computer Arithmetic: Design and Implementation*, New York: McGraw-Hill, 1986.

[14] H.B. Bakoglu. *Circuits, Interconnections, and Packaging for VLSI*. 1990. Massachusetts: Addison-Wesley Publishing Company.

[15] H.W. Ott. *Noise Reduction Techniques in Electronic Systems*. 2nd. Ed., 1988. New York: John Wiley Interscience.

[16] Larry D. Smith, 'Decoupling Capacitor Calculations for CMOS Circuits', *Proc. of the Electrical Performance of Packages Conference*, 1994, pp. 101–105.

[17] Jong-Gwan Yook, V. Chandramouli, Linda P. B. Katehi, Karem A. Sakallah, Tawfik R. Arabi, and Tim A. Schreyer, 'Computation of Switching Noise in Printed Circuit Boards', *IEEE Tr. on Components, Packaging and Manufacturing Technology—Part A*, Vol. 20, No. 1, March 1997, pp. 64–75.

[18] Bob Mammano, 'Fueling the Megaprocessors-Empowering Dynamic Energy Management',

[19] Howard H. Chen and Stanley E. Schuster, 'On-Chip Decoupling Capacitor Optimization for High-Performance VLSI Design', *Proc. 1995 IEEE Int. Symp. VLSI Tech., Systems and Appl.*, pp. 99–103.

[20] Balsha R. Stanisic, Rob. A. Rutenbar, and L. Richard Carley. *Synthesis of Power Distribution to Manage Signal Integrity in Mixed-Signal ICs*. 1996. Massachusetts: Kluwer Academic Publishers.

[21] Robert C. Frye, 'Passive Components in Electronic Applications: Requirements and Prospects for Integration', *Int Journal of Microcircuits and Electronic Packaging*, Vol. 19, No. 4, 1996, pp. 483–490.

[22] A. Chamas et al., 'A 64b Microprocessor with Multimedia Support', *1995 IEEE Int. Solid-State Circuits Conference Digest of Technical Papers*, February 1995, pp. 178–179, 361.

[23] Chender Huang, Yaochao Yang, and John L. Prince, 'A Simultaneous Switching Noise Design Algorithm for Leadframe Packages with or without Ground Plane', *IEEE Tr. on Components, Packaging, and Manufacturing Techn.—Part B*, Vol. 19, No. 1, February 1996, pp. 15–22.

[24] John M. Williamson, Michel S. Nakhla, Qi-Jun Zhang, and Patrick D. van der Puije, 'Ground Noise Minimizatin in Integrated Circuit Packages through Pin Assignment Optimization', *IEEE Tr. on Components, Packaging, and Manufacturing Techn.—Part B*, Vol. 19, No. 2, May 1996, pp. 361–371.

[25] Shaofang Gong, Hans Hentzell, Sven-Tuve Persson, Hjalmar Hesselbom, Bo Lofstedt, and Magnus Hansen, 'Techniques for Reducing Switching Noise in High Speed Digital Systems', *Proc. 8th Annual IEEE INt. ASIC Conf. and Exhibit*, 1995, pp. 21–24.

[26] 'Ground Bounce in 8-Bit High Speed Logic', Harris Semiconductor Application Note No. AN9646, December 1996. http://www.semic.harris.com

[27] Carmen Mattei, Bruce Gueinin, Phil Wingate, and Richard Ried, 'Considerations for Designing and Utilizing Plastic MCM BGA Packaging', *Advancing Microelectronics*, Vol. 24, No. 2, March/April 1997, pp. 20–25.

[28] Kumaresh Mathey, Madhavan Swaminathan, L. D. Smith, and T. J. Cockerill, 'Noise Computation in Single Chip Packages', *IEEE Tr. on Components, Packaging, and Manufacturing Techn.—Part B*, Vol. 19, No. 2, May 1996, pp. 350–360.

[29] P. Vanoostende, et al., 'Evaluations of the Limitations of the Simple CMOS Power Estimation Formula: Comparison with Accurate Estimation', *PATMOS*, 1992, pp. 16–25.

[30] D. Liu, C. Svensson, 'Power Consumption Estimation in CMOS VLSI Chips', *IEEE J. Solid-State Circuits*, Vol. 29, No. 6, June 1994, pp. 663–670.

[31] Patrik Larsson, 'Analog Phenomena in Digital Circuits', *Linköping Studies in Science and Technology*, Dissertation No. 376, 1995.

[32] Michael Dolle, 'Analysis of Simultaneous Switching Noise', *IEEE Symp. Circuits and Systems*, 1995, pp. 904–907.

[33] Samil Hasan and J.L. Prince, 'The Effect of Core-Logic Parasitic Capacity on the Power Bus Noise', ECE 696B Report, Center for Electronic Packaging Research, Dept. of Electrical and Computer Engineering, Univ. of Arizona, Tucson.

[34] R. Downing, P. Gebler, and George Katopis, 'Decoupling Capacitor Effects on Switching Noise', *IEEE Tr. on Components, Hybrids, and Manufacturing Technology*, Vol. 16, No. 5, August 1993, pp. 484–489.

[35] Zhonghua Wu, Yuzhe Chen, and Jiayuan Fang, 'Modeling and Simulation of Integral Decoupling Capacitors in Single and Multichip Module Electronics Packaging', *Proc. 44th Electronic Components Technol. Conf.*, 1994, pp. 945–948.

[36] C.C. Huang, Bill Loh, and Florence Wong, 'Ground Bounce Study of 304 Lead Interposer MQFP with On-Chip Decoupling Capacitor Test Die', *Proc. IEEE Tech. Applications Conf. NORTCON'95*, 1995, pp. 343–346.

[37] W. Bowhill, et al., 'A 300 MHz 64b Quad-Issue CMOS RISC Microprocessor', *Proc. Int. Solid-State Conference*, February 1995, pp. 182–183.

[38] Liren Chen, and Bidyut Sen, 'Measurement Study on Simultaneous Switching Noise', *Proc. 4th Topical Meeting on Electrical Performance of Electronic Packages*, 1995, pp. 40–42.

[39] Bidyut K. Sen, and Richard L. Wheeler, 'Performance Comparison of Discrete and Buried Capacitors', *Int. Journal of Microcircuits and Electronic Packaging*, Vol. 19, No. 4, 1996, pp. 449–455.

[40] Mark Ingels, and Michiel S.J. Steyaert, 'Design Strategies and Decoupling Techniques for Reducing the Effects of Electrical Interference in Mixed-Mode IC's', *IEEE J. of Solid-State Circuits*, Vol. 32, No. 7, July 1997, pp. 1136–1141.

[41] J.L. González, and A.Rubio, 'TCMOS: Low noise power supply technique for digital IC's', *Electronics Letters*, Vol. 31, No. 16, pp. 1338–1339, 1995.

[42] Disaburo Takashima, Yukihito Oowaki, Shigeyoshi Watanabe, and Kazunori Ohuci, 'Noise Suppression Scheme for Gigabit-Scale and Gigabyte/s Data-Rate LSI's', *IEEE J. Solid-State Circuits*, Vol. 33, No. 2, pp. 260–267, February 1998.

10 Conclusions

With the trends now found in electronic technology, increasing complexity, supply voltage reduction and very high switching speed, parasitic coupling between circuits is becoming a critical issue in the design and testing of modern and future integrated circuits. Inductive, capacitive, resistive and electromagnetic coupling mechanisms are limiting the performance possibilities of VLSI circuits, meaning that design engineer's understanding of such principles is one of today's major issues in both academic and industrial sphere.

One way to classify interactions is provided by the source location's relation with the receptive circuit. Long distance interactions, in terms of VLSI circuits' coupling between different integrated circuits in the same printed circuit board or cabinet, are due principally to electromagnetic interference (EMI) and voltage distribution transients through the power supply bars and lines. As regards short distance interactions inside the same integrated circuits, packaging and layout parasitic components, substrate, interconnection coupling, thermal coupling and EMI effects again can be considered. The behavior and shape of currents in modern circuits (CMOS and BICMOS) displays sharp peaks (high level dI/dt) that cause undesired voltages (switching noise voltage) through inductive parasitic coupling (packaging, layout). Complementarily, together with this effect, the increasingly wider and higher frequency band of signals in circuits, especially in digital circuits, favors coupling through capacitive parasitic components (substrate and interconnection lines).

Though negligible for previous technology generations, such topics are critical for today's high performance circuits, and there is consequently a need for modeling investigation in order to cover that principles by modern CAD systems as well as for fostering suitable skills in mixed-signal design training courses. This book has focused its attention on two closely connected mechanisms, substrate coupling and switching noise, both critical in the microelectronic design of today's integrated circuits. These are also connected with packaging aspects which, though considered in the discussions, were regarded as being outside the scope of the book.

10.1 Conclusions as regards substrate coupling mechanism

Substrate is the general means that allocates all the devices in an integrated circuit. It is usually the basis for the creation, through the different phases of process fabrication, of the PMOS and NMOS transistors forming the circuit. It is thus clear that the substrate becomes a channel through which electrical, mechanical and thermal interactions are transmitted. From the standpoint of the first of these, the different semiconductor parts of components are insulated from each other because of potential barriers caused by the reverse biasing of junctions. These barriers nevertheless display conductive behavior for dynamic signals by coupling the different nodes of the circuits to the distributed resistive mesh of the substrate.

Assuming the possibility of differentiating the noise sources from noise receivers, the chain of interaction can be structured as injection mechanisms, coupling means and reception mechanisms. Two major injection mechanisms can be considered: the one caused by reverse-biased junction (drain and source to substrate or well) and the one caused by the direct connection of noisy lines (usually *Vdd* and *GND* lines) to substrate (contacts). The substrate or bulk is the coupling means and can be modeled as a mesh of three-dimensional distributed resistances and capacitances. Finally, the major reception mechanisms of substrate noise are substrate contacts, and reverse-biased junctions that affect both the drain/source terminals of devices and the channel of the MOS transistor (body effect).

The way noise is coupled and propagated, in terms of a global integrated circuit, depends on layout and technology parameters. As layout (or design) factors, the distance between devices, the components and contacts present in the path from source to receiver, the rise/fall time of signals and the bias point of the sensitive devices should all be taken into account. As technology factors, the substrate type (P-, P+, SOI), the wafer thickness and the epitaxial layer thickness have to be considered. Chapter 2 has evaluated the weight of all these factors by both simulation and experimental measurements. Also, in this chapter it has been seen that coupling is minimum when devices are in different wells. Thus, the junction provided by a well is also a good means to obtain insulation.

Coupling drops linearly with distance in lightly doped wafers (P-) depending on the impedance of the return paths encountered by noise along its propagation. For a typical technology, the noise attenuation ratio is two orders of attenuation per 100 μm. In the opposite sense, highly doped wafers (P+) do not necessarily drop with distance, but the highly conductive bulk behaves electrically as a single node and noise is therefore eventually constant. Because of the isolating characteristics of SOI technologies they can be considered as an interesting and promising technique for high performance mixed circuits.

The relative noise contribution of bipolar and MOS devices in a BiCMOS technology was investigated through both simulation and experimental measurements. The contribution of BJT devices in such a technology is one order higher than MOS devices. Finally it has been proved that a fine bulk biasing contact distribution layout is one of the most effective mechanisms for diminishing noise coupling, this being the basis of the guard ring concepts.

A guard ring is a low impedance tie in the disturbance path, providing a return path to *GND* or *Vdd* and thus preventing noise from reaching the sensitive circuitry. The analysis of substrate noise propagation shows the effectiveness of guard rings, sets of contacts biasing substrate to low impedance sources, located in the path between noise and receiver or surrounding one or both sections. The distance between contacts factor is connected with technology parameters. For ideal conditions it has been shown that guard ring techniques can be classified from least to most efficient as follows: oxide trench, N-ring to *Vdd*, N-ring to *GND*, P-ring to *GND* and buried layer. It is concluded that the guard ring must be of the same type as the substrate, as implants of other types have proved to be inefficient.

For the usual wafer conductivity and thickness, skin effect appears at frequencies around a gigahertz, and at this or higher frequencies two significant conclusions can be drawn. First, the effect of backside contacts is practically eliminated, resulting in worse substrate biasing. Second, in heavily doped substrates the frequency at which this regime is entered is low enough to consider that the beneficial effects of a backside contact may in practice be reduced or even eliminated.

Mixed signal integrated circuits, where both fast digital and sensitive analog parts are integrated together, are the most critical scenario in which coupled noise has to be taken into account. The two main ways through which noise can be coupled to an analog node are the lines supplying power to the circuit, and the substrate mainly because of body effect. Noise propagation is attenuated with distance in lightly doped substrates. Consequently, keeping a safe distance between the noisy digital part and the analog section could in principle be enough to minimize or control the effects of substrate noise. The value of this distance will depend on the level of noise produced by the digital section and the noise tolerance of the analog section. However, the benefits of distance may be lost if a single line biases all the contacts in the IC, since this technique short-circuits the substrate of both sections. Biasing all the substrate only with the analog or only with the digital ground should be consequently discarded. On the contrary, the analog and digital contacts should be placed far enough apart to prevent noise injection.

The biasing strategies minimizing noise coupling are those biasing the substrate at the digital section with digital ground, and the substrate at the analog section with an analog ground, or even better a dedicated supply. The first one minimizes body effect but favors noise coupling through contacts. The second one offers in general the lowest noise levels, although it needs an extra supply line, with the consequent increase of area, routing complexity and package pin number. Measurements in a test circuit show that the analog supply lines should in practice never be used to bias the substrate, because noise is easily coupled through them, even if contacts are located at the pad periphery. On the other side, biasing the digital section with a line other than the digital ground may induce latch-up problems. This is an inconvenience because supply lines are usually contaminated with high switching noise, which will then be introduced to the substrate. Thus first aim will be to reduce switching noise at supply lines as much as possible, either with circuit or architecture techniques as discussed in the last chapters of the book, either with packaging techniques. In fact it has been proved that the contribution of the power supply package pin inductances is larger than what is caused by biasing policy.

In heavily doped epi wafers the behavior of noise propagation and biasing are different. The highly conductive part of the substrate allows a very low impedance path between all contacts and transistors, so distance is no longer a factor to be taken into account. If supply lines were connected through an ideal package, biasing would be perfect, all noise would return to ground to ground and noise coupling would be negligible. Nevertheless this does not occur in practice because of the effect of package parasitics, and return is partially through other devices. Even backside contacts are not efficient for eliminating noise. Given the usual size of noisy sections, the values of pin inductance necessary to allow good return paths should be very low, under 1nH, which is difficult to achieve even with multiple assignment. So, from a noise standpoint, the use of P+ wafers is not advisable for mixed signal circuits. Measurement results show noise levels about three times larger in P+ wafers as compared to P- ones.

As a complement to the (passive) guard ring technique for reducing noise coupling as discussed above, other more sophisticated techniques can be applied. As the most relevant of these, buried layer techniques, noise compensation techniques and active guard bands can be considered.

Buried layers are highly conductive layers under the switching devices providing a low impedance path for noise derivation to ground. The results show that it is more efficient to place the buried layer under the noisy circuitry than under the sensitive part. In a case where the level of noise without buried layer gave a level of about 250 mV, it was possible to reduce noise one order with deep bias contacts and buried layer under the whole circuitry. Although a buried layer implies a significant cost increase in CMOS technology, it does not mean an increase in process complexity in BiCMOS technology.

The second complementary technique is an active one. The idea is to cancel noise caused by a switching device with the noise caused by another device switching in a complementary sense. This idea may be used at device level or higher, allocating complementary sections symmetrically in relation to the sensitive components. Full cancellation, though theoretically possible, is not attainable in practice. Small differences in load, placement, and timing cause a limitation. This technique is closely connected with a promising topic of Design for Low Level of Noise (DLLN). The third method, active guard bands or rings, is connected with this second technique but does not involve the need for a formal complementary. In this technique, a fast feedback differential device (like an OPAMP) minimizes the signal appearing in the differential input, where one input terminal is connected to the substrate and the other to a high quality ground. This technique has shown to be efficient for a low frequency noise spectrum but is very limited for very fast transient noise (digital switching).

10.2 Conclusions as regards switching noise

Switching noise, also known as delta-I noise (*dI/dt*), arises through the undesired appearance of voltage spikes across parasitic inductance (layout and specially packaging) due to the abrupt shape of power supply currents in CMOS and BICMOS technologies. These two technologies are not significantly different in terms of delta-I noise. The global phenomenon is also known as ground bounce. A special effect appears in the case of fast wide-word digital processors where nodes (especially external bus drivers) may switch simultaneously (simultaneous switching noise, SSN). This noise is a very limiting one in today's advanced processors. The noise caused by an increasing number of simultaneous switching gates is not proportional to the number of gates but increases sublinearly.

Two basic techniques can be considered to reduce switching noise: controlling parasitic capacitances (low *L* packages, multiple pins and pads, low *L* internal routing) and avoiding simultaneous switching of drivers by introducing adaptive delays (this technique is found in many commercial processors). This fact implies that designers have to take into account packaging (incorporation of advanced models for packages in design, simulation, validation and testing) and advanced extractors. The designer also has to take design parameters and strategies into account in order to get a low level of noise without functional performance limitation. The different factors that the designer may take into account are, at logic level, the fanout levels of the logic, the parallelism level, the coding of variables and in general the number of simultaneous switches. At transistor level these are the supply voltage levels, the transistor sizing and the output loads.

Delta-I noise is closely connected with substrate coupling mechanism. Delta-I noise generated usually from packaging parasitic inductance is very likely to be injected to the substrate through biasing contacts. Although digital static CMOS circuits are very resistant to switching noise this is not the case for other digital circuitry as ECL, TTL and especially dynamic CMOS. The effects that switching noise may cause in digital circuits are transient errors, read/write faults, violation of codes, changes of state of the machines, with a possibility of device failures appearing due to over-voltages. This fact has recently led the design and testing community to considering noise as a source of faults.

As regards the analog section of a mixed signal integrated circuit, switching noise produces a limitation in performance by reducing the dynamic range of the OPAMPs or introducing phase errors and degrading the SNR in the input stage of the ADCs, thus limiting the resolution possible.

High level languages are designed for getting a compact description of very complex systems and for acting as the input of automatic synthesizing tools in high productive design environments. Synthesizers tend to have two different goals: minimum area and maximum performance (speed). These strategies fail to consider low switching level. The former is nevertheless best suited for low noise design, and additional rules may be applied at high level to reduce SSN, such as re-timing data-flow graphs to minimize the number of operations performed at the same time.

At structural level, two-level logic is prone to generate high levels of switching noise. Different multilevel structures for the same function display different levels of switching noise. Many of the techniques used in digital design for obtaining low power may be suitable for low noise, but not all of these are. Clock gating provides low noise but pipeline and concurrent processing increase noise levels. Even several low power techniques like the power-down modes are harmful in terms of noise.

Conventional static CMOS logic gives a good compromise between noise generation and noise margins, but other alternatives (self-timed clocked CMOS and low output voltage swings) involves better merit factors.

Three sophisticated techniques for digital design with low noise levels, different from the basic, were presented in the previous chapter. The first involves the use of logic with different working principles from those existing nowadays and providing constant current consumption (cancellation of delta-I noise). The most representative of these could be CSL, ESCL and FSCL logics, that reduce switching noise but entail an important drawback, the significant increase in power consumption. A second technique is the use of CMOS logic gates but with a design principle that leads the overall consumption to minimize *dI/dt*. This technique, which has been shown experimentally, is based on the aggregation of trapezoidal time/consumption pieces in order to get a quasi-constant summation. Contrary to the previous techniques, this one does not increase the circuit's power consumption. The drawback of this technique is that it is necessary to have a uniform diffusion of signals, independent of the input combination, and to obtain such diffusion a well-structured circuit that tends to have a moderate or medium gate overhead is essential.

Finally, the third technique deals with the general concept of implementing all or part of the power supply in order to minimize or eliminate *dI/dt* through the packaging. Amongst many others presented in recent publications, the TCMOS technique was expounded. Intermediate circuitry inside the packaging located between the *Vdd/GND* terminals and the conventional CMOS logic circuit reduces *dI/dt* activity by feeding the logic from a buffer circuit that stores energy in an on-chip capacitance.

MIX
Papier aus verantwortungsvollen Quellen
Paper from responsible sources
FSC® C105338

If you have any concerns about our products,
you can contact us on
ProductSafety@springernature.com

In case Publisher is established outside the EU,
the EU authorized representative is:
Springer Nature Customer Service Center GmbH
Europaplatz 3, 69115 Heidelberg, Germany

Printed by Libri Plureos GmbH
in Hamburg, Germany